Owning the Keys to Discover and Open Mysteries of the Universe

Joyce Morford

First Edition- Series1

Published 2016

Book layout by www.ebooklaunch.com

Table of Contents

ABOUT THE AUTHOR

Joyce Ann Morford came to Earth on an April Day not knowing that Self Realization in a world of Duality was to be her path. Joyce Ann Morford became a Spiritual Lighthouse for those that would listen.

Joyce Ann Morford's Possesses No PHD, no Master's degree.no Bachelor's Degree, or any such degree.

Joyce Ann Morford's credential's come from the inside out to share with all Humanity.

Joyce Ann Morford is a natural born psychic medium handed down from lineage. Joyce Ann Morford walked in a Christ Conscious State; or you can call it a Quantum state of consciousness still attached to the physical body for over 15 years, not needing a PHD, or a Master's degree, or a Bachelor's degree to obtain it. She did not need to be one of the "Elite" nor did she need to be a famous person to obtain it.

Joyce Ann Morford realizes she is a spirit in a body; in touch with her Monad, having a Human experience and her Name made her that here on Earth! My name and the entanglement of other's name's created my experiences. Each one unique with that person.

Joyce Ann Morford realizes that "I am You and You are Me! We are One, One, One, now and forever Amen!

ACKNOWLEDGEMENTS

I would like to thank the "One Infinite Creator" for the many ME's I have been which helped me to develop my inner senses of Clairvoyance, Clairaudience, Clair Sentinance and Claircognizant abilities. Without you I would not have existed at all. I Love You

I would like to thank my Monad for looking back upon itself in each lifetime guiding me to where I am today. I thank you for my Awakening to your presence and guidance throughout this life time. I thank you for bestowing your loving light upon me and giving me the realization that all is temporary here on Earth and only an Illusion. I Love You.

I want to thank everyone on the other side of the veil for helping me see the truth in all matters. For teaching me and holding my hand, but most of all for helping me help others in need of answers. I Love You

I want to thank my grandmother on the other side of the veil for staying with me throughout my lifetime as a wonderful Mother looking out after me, Loving me, also giving me guidance and answers so that I may be safe and walk the right path. I can't wait to see you on the other side! I miss you. I Love You

I want to thank my twin sister Judy and my Brother n law for giving me a safe haven to live. For without their help this book would have never made it to fruition. I Love You

I want to thank my dog Jupiter for his companionship and unconditional Love. He truly is my best Friend. I Love You

I want to thank everyone who gave me a kind ear, a smile and a hug when I needed it; Especially the ones that gave me encouragement to be myself; and also to not back down from nay-sayers.

I want to thank all the darkness in my life; for without the darkness I would not have received the gifts of Strength, Courage and Resolve. I would not have had the ability to stand up to you! I Love You

A Message to all Humanity, "Love is a Universal Solvent that dissolves all Paradoxes". My Hope is one day all of Humanity will raise their Consciousness to Love and Oneness so we can have peace on Earth. To end a Duality Consciousness that is steeped in separation. The separation is an illusion; don't wait to be on the other side to have this realization. I am You and You are me forever and ever! I Love You

PRELUDE

METAPHYSICS- Relating to a reality beyond what is acceptable to the senses. Of or relating to things that are thought to exist but cannot be seen. Some would call it supernatural.

This is a book of a Metaphysical Treatise and a Metaphysical experience.

This is a book about my own journey; a book about Awakening to my own inherent abilities as a Spiritual being having a Human experience.

I am that I am presence is a powerful realization in this lifetime on Earth for me.

Today I am a light worker who has awakened a powerful spiritual gift. The gift that I am sharing with Humanity is the gift of Nomancy. Nomancy is the gift of: divining the destiny of a person, places or things by the letters in their name. The gift that I have is a rare psychic gift. The names you have in this lifetime were chosen by you; it was no accident! We chose the vibration of our names in order to help fulfill our mission here on Earth. Your mission could be for Soul growth, or it could be for karmic lessons to learn in this lifetime. Your

Soul mission could entail finding the God within yourself. The Acts, The Scenes, The Story were contracted with other souls on the other side of the veil before this incarnation.

My purpose as the Seer: A Seer, or Psychic or as I like to call myself an intuitive empath, is a person who has their ego set aside, hence they are Egoless. In that state of consciousness one can read the person's Ego-Personality and through the name one is able to see the Act, the Scenes, The Story, of the life on Earth. The Name is the same as D.N.A. The Seer, the Psychics, The Intuitive readers are all reading D.N.A.; Your D.N.A.

The story in this book is about my personal journey towards my Spiritual Evolution and the trial and tribulations I went through to get there. The trials and tribulations were chosen by me. Without these things I would not have awakened my Soul to my life's purpose on Earth. I am here today as an Awakened Spiritually Evolved Soul, to relay Spiritual Truth to Humanity.

My divine mission in this life time is to relay new concepts for humanity. New knowledge to share with Humanity. Namaste to all.

YOUR NAME TRUTH

"A Name is the Same as D.N.A."

INTRODUCTION

I am writing this book in hopes to help humanity in some way in this life time.

I am writing this book because my Consciousness went through an initiation and I stayed in a Christ Consciousness state for over 15 years. My life will never be the same; but at the same time my shift in Consciousness will help the rest of humanities Consciousness to shift also.

I am writing this book due to a major experience I had starting in the year 2000. Before I elaborate on the experience I am going to give a little background information about my self.

I guess I will say I am an average person living an everyday average life; meaning I do not possess a P.H.D

Nor do I possess any kind of higher credentials.

What I do possess is a Clairvoyant, (clear seeing) Clairaudient, (clear hearing), Clair cognizance (knowing)

Clairsentinance (feeling), I have been a sensitive person all my life with the feeling of not belonging here. My interest'

Has always been about life and why I am here. I have always been fascinated with Metaphysics, Esoteric Knowledge, UFO's, the Supernatural, and Life after Death, etc.

I guess the correct term would be New Age subjects.

My experience occurred when I had turned 41 years of age in 2001. I was married only for a short while. I went through a Kundalini (spiritual orgasm) experience affecting and clearing all of my chakras. What is kundalini? Most people of today do not have any idea what a kundalini awakening is. The people who do know are those who are probably on a spiritual path. Kundalini is the Serpent force that lies coiled at the base of the spine. It is most often awakened in physical sex and orgasm. However; in the spiritually minded enlightened Man or Woman the kundalini serpent undulates up through the spine opening each of the seven chakras on the way. Once it reaches the Pineal Gland the spiritual eye is opened and man/woman comes into conscious contact with higher realms. You come in contact with your Higher Self.; a Direct communication with a higher aspect of yourself. I call it your Christ Self; it is Christ Consciousness! It is your EGO-less self! My body was vibrating at a very rapid pace, and my consciousness expanded 3 fold. The most profound aspect of it all was the loss of my EGO (persona or personality) and my Real Spiritual Self was front and center. At that moment I was able to communicate with my Higher Self. Your Higher Self resides on the inner planes with-in the body. It resides in a higher dimension than 3d Earth. Your light here on earth is only a small portion of your self-animating the physical body you currently reside in. Your physical body would not be able to handle the energy of your whole self; it would explode! In order to communicate with a larger portion of your self is to go with-in be still; be quiet; and silence the mind. Meditation is the key. Being in a state of balanced male and female energy in the body; everything was so serene and peaceful. Everything was bliss. Everything was Oneness! There is no Polarity, no Duality. You are one with everyone, and everything! You are as close to being God that you can ever experience in this state of one-ness. In Reality there is no separation. Only your false self the Ego is separation! I felt this pressing need to help all of humanity in some way.

I found that this process of natural evolution is not being taught to humanity. It has been kept as esoteric doctrine for millions of years by religious entities alone. I find it is the biggest injustice to humanity. As Jesus once said "The kingdom of God is within you." You are the living God. Your temple is your body. Your consciousness is eternal.

I find it a disgrace to Humanity; that they are told that one's redemption can only come through a church.

I found that raising the Christ Seed within your self is your redemption. For as you raise this Christ Consciousness you are born a new, you are as white as snow. The church tells you not to go within yourself. Going within yourself is the only way to becoming the Christ Soul that you are. The seeker must go within to find the truth! To know itself to be itself; it is the only way. Meditation within is your salvation. Here is a quote from Edgar Cayce "All should be taught in the beginning that we are one, one, and one.

I believe in this New Earth; the energy of kundalini awakening will be happening to others quite often. It is the awakening to the truth of your-self. Your spiritual evolution is raising the energy around your body called the Merkaba. This is a male and female energy; your light body. This energy field has geometry to it. While I was in Christ Consciousness I kept drawing the Seal of Solomon -the Star of David not knowing why until I channeled the name.

NAMES AND QUANTUM ENTANGLEMENT

Quantum Entanglement can be seen by the interaction on the material plane of names and the reality it produces in the physical world the manifested plane.

The names arts I introduce in this metaphysical treatise is done objectively without the ego involved.

NAME= ME A NAME AMEN
AMEN= ME A NAME AMEN SEE THE MIRROR OF NAME AND AMEN

The name arts or Science introduced in this Metaphysical Treatise shows Quantum Entanglement on Earth with others. It can be seen as a whole group of souls under one vibration: those living in the United States of America for example. It can be seen in your own birth name by itself. It can be seen when you marry or date someone. It can be seen with your relationships of your mother and father, sister and brothers; even co-workers and friends. Each entanglement will give you the acts the scene's and the story of each relationship here on Earth. After you have read the name arts objectively, you can see the illusion in it all. Every name entanglement reveals the acts; the scenes and the story of all. Everyone is a piece of the whole with a story to tell. The real test is waking up and seeing it as a story a game for the one Creator to experience itself. Even the tragic events are the one creator experiencing it through you a co-creator.

Karmic ties are acted out on the material plane for balancing and soul growth the karma you reap what you sow in the forgetfulness of your real self.

The Ego; the false self does not care what it does to another when unleashing its negative traits. For the ego self believes in its false self. The ego hides the true nature of the soul in the body.

If one can re-mem-ber the soul in yourself and see the same in others as a mirror to self; then perhaps you will go beyond the ego self and do rightly by the soul self in you and others.

For the ultimate power is "LOVE" and God is "LOVE" God is benevolent' all human atrocities on Earth have been done to humans by the false self the Ego. Man does these evil things It is not of God. "Love hurts no one".

A consciousness raised to a quantum state annihilates the Ego and any separation, the ego cannot survive there. All is Joy, Love, Oneness and

Peace. A higher consciousness is what everyone should strive for.

Meditation within oneself is a way to diminish the false ego self so your true light will shine forth.

MY INITIATION INTO CHRIST CONSCIOUSNESS

The birthing of Christ within you.

It started with my extensive search for the meaning of life.

I searched every esoteric book I could find. I read everything that I could get my hands on. During the period I would occasionally see other psychic's for readings. I was told to keep reading; to keep doing my spiritual practices, that's all I was told for quite some time.

One night I started having very vivid dreams and spontaneous orgasms in the middle of the night. As I would have these orgasms I could feel energy moving up my spine. It was a slow process and also quite uncomfortable sometimes. I found myself sleeping with a heating pad; it seemed to help the aching pain I was having.

The first dream I had I found myself going through a wormhole and then a tunnel; there was a well and then what appeared to me to be a black sun.

The second dream I had was an explosion of a planet. As I researched the Black Sun and the explosion of a planet I came across the Planet Malduk.

It was dark; I saw a door and I was fumbling around frantically trying to find the key to the door. I felt I had to open the door!

I decided to go back to my psychic to see if she picked up on any of this. I felt I was being guided from the inside. My psychic told me I had a Spiritual Teacher with me; and to meditate more often. She had given me confirmation as to what I felt myself.

I was told I was psychic and it was from my lineage.

I was aware of my Psychic Gifts, although sometimes I just needed verification from another psychic.

As time went on I started to be awakened at 3:00 Am and I would be guided to different website on my computer pertaining to Geometry and the Giza Pyramid. Night after night for a few weeks I was gathering all I could on the subject.

The third dream I had was a dream of Thoth the ancient scribe.

Now the pains I were having at night became more intense along my spinal column, and my heating pad was now between my shoulder blades to the base of my neck. Sometimes it felt like my neck was just going to snap. Again I was up at 3:00 Am and now I was researching Thoth. I started reading the Emerald tablets of Thoth trying to discern the information.

The fourth Dream I had was of a being named RA. He was very tall with a human body and a bird face that was blue in color. I remember calling him Amon-Ra, and I started seeing Egyptian Iconography in my dreams; the pyramids' the left and right eye of Horus. I saw myself in a coffin and then the lid was shut.

At this point I was guided to read "The Law of One"

The pains I was having were really intense around my neck, especially at the base of my neck where the spine ends. (At this point I saw a ladder presumably at the time I figured it was my D.N. A). My D.N.A. was lit up and swirling around like 2 entangled ladders with very vivid colors as in the chakras and crystal clear brightness. I realized it was a kundalini awakening all my chakras were opened.

The Fifth Dream I had was that of the Pyramids, the EYE of Horus; and Quazecoatal.

At this point I had so much energy within me that my body was shaking. My mind (consciousness) was separated from my body. I realized that my consciousness was encased in my causal body within my star tetrahedron's rotating very fast. I was riding in my light vehicle. I was in phase lock with the Giza Pyramid. My consciousness started to move up through the Pyramid; as it moved upwards I saw the face of Jesus in my third eye. I did not know the meaning of its significance until 10 years later While reading "The Flower of Life". My Consciousness continued upwards where it exited out the top of the pyramid and my consciousness was soaring into space enclosed in my merkaba, my star car. My consciousness was now in the center of the Milky Way and I gasped at how beautiful it was and that I wanted to stay. I was in such a loving, peaceful place that I did not want to leave. I left this experience in the Christ conscious grid for humanity. My consciousness said I was to return to help humanity. (It is like the hundredth monkey syndrome which is as more and more of us make this ascension process it will enable more of humanity's D.N.A. to spontaneously to be able to do this.)As my consciousness was returning back to earth; I found myself looking at the coffin my body was in. Then lying next to that coffin was another one and my consciousness entered the new coffin. I saw after my consciousness entered the second coffin both coffins flew open and my consciousness was changed from the old to the new consciousness. My consciousness was in a state of oneness, joy, peace and belonging. There was no separation with anyone or anything; all was connected! The birds were singing. I was in total bliss and happiness. It was a resurrection for me. Then I saw the face of Jesus in the pyramid. I was in a Christ Conscious state. I stayed in this state of consciousness for over 15 years. My life would never be the same after having the experience of being in a Christ Conscious state. The energy of my body was at a very high vibration, all my chakras were open and I found myself communicating Multi-dimensionally with other beings. All my psychic centers were open. When I would look at a name, any name I am able to see the acts; the scenes; the story of that entities life. There was no duality, no Judgement, when I read a name, I was neutral, I am looking at the names objectively. I was one with the whole of humanity! I possessed no EGO!

It wasn't until years later that I realized what I went through. I feel this is what I am to leave for humanity; my experience into Christ Consciousness and living it for over 15 years. With the hope that humanity will understand that Christ is manifested within you, Just as Jesus manifested the Christ within himself. Christ Consciousness is an Initiation in Consciousness; it is not Jesus returning; so many people are waiting for his second coming. Christ cometh within you! Christ suffers all human experience with in you! Christ is the Comforter!

Note: In my state of Christ Consciousness I drew the symbol of The Seal of Solomon; The Star of David all the time. This is the Christ Star! This is your light vehicle the merkaba. This symbol was the state of Consciousness I was in. The union of opposites.

A QUEST FOR ANSWERS

As this phenomenon kept happening, I started to investigate everything from the Bible, History, or other Authors whom might have mentioned it. Intuition and Synchronicity has always been a big part of my life, and I have never been one to discount it. I use these tools quite often and I am often guided to the right information that my mind has been pondering on.

When my Gift and my purpose in this life appeared, I was guided to the information that I seek.

Quote: Edgar Cayce Companion Page 172

There is a Name to each Soul! For He hath called His own by Name! What Name? All the Names that may have been as a Material experience through the experience of an entity in that environ or those relative associations of Venus, Mars, Jupiter, Uranus, Saturn, Sun, Moon, or what. Or a Name that is above every Name. Every influence you see is relative! Hence the Name is relative to that which is accomplished by the soul in its sojourn throughout its whole experience. Was one named John by chance? No! Was one Named Joe or Llewellyn by chance? No; they are all relative! In the end the name is the sum total of what the soul entity in all its vibratory forces has borne toward the Creative Force itself. Names have their meaning, and these depend upon the purposes when such are bestowed upon an individual entity entering the Earths plane.

Have ye not understood how that in various experiences individuals, as their purposes or attitudes or desires were indicated, had their names henceforth called a complete or full name meaning or indicating the purpose to which the individual entity or souls had been called?

Question: When would it be best to choose a name for a child?

Answer: When ye have determined as to the purpose to which ye hope, and which ye will, which ye are willing to dedicate same.

Names have their vibration; to be sure, names have their element of Influences or Force, by the very Activity of the name.

Question: How would an entity benefit by changing its name?

Answer: How did Abraham benefit by changing the name from Abram to Abraham? How did Paul gain by the change from Saul to Paul? How was there difference in the names of the Hebrew children in their varied surroundings? There is builded upon each name that which carries its own meaning or significance, that gives rather the impelling of, and the lifting up to, meeting such conditions. So does the changing or altering of a name set about varied environs or vibrations that make for the conduciveness of changed surroundings?

It is given in thy writings of Scripture, although in a HIDDEN manner, ye may observe if ye will look how Adam named those that were brought before him in creation. Their NAME indicates you the carnal mind their relationships in the sex condition or question. Each animal, each bird, each fowl, has been so named for some peculiarity of that individual beast, bird or fowl, and in this manner represents some particular phase of man's development in the earths plane, or that consciousness of some particular element or personality that is manifested in man.

In all of the animals... on their natural state these are in the FORMS as their NAMES indicate. And from these man may learn many lessons; which was attempted in the beginning.

Who having named the Name of Christ has become conscious of that He represented or presented in the world?

As the records have been handed down that Abraham represents the faithful, Moses meekness,

David the warrior yet humility, so the Christ represents LOVE!

When each entity, each Soul has so manifested, so acted in its relationships as to become then as the new Name; white clear, known only to him that hath overcome. Overcome what?

The World even as he; as has been indicated in another portion of Revelation, all those that bear the mark, those that have the name, they have the stone... they have the new Name, they have the Manna, they have the understanding; As to Times; as to Seasons, as to Places. Alone it is given to those who have named the name and who bear the mark of those of His calling and His election in their BODIES. To them it shall be given.

Ye shall know Him even as He knows thee; for He calls thee by Name.

Hence they that have the Name of the Son have put their trust in Him. He is their standard, their model, their hope, their activity.

"The Law of One" Original channeled by Carla Reukoff RA states: All names can be taken to be created for the vibrational characteristics. If one wishes to have power over an entity it is an aid to know that entities name.

If we wish no power over an entity, but wishes to collect that entity into the very heart of one's own being it is well to forget the name. Both processes are viable magically.

Quote: I am Ra: "The Law of One" though Beyond the limitations of "Name" All things are one- no polarity-no righteousness-no disharmony, but only the identity "all is one" love/light-light/love.

QUEST 2
THE SEARCH CONTINUE

Edgar Cayce.org Biblical Reference: Book of Life: Akashic Records: Names:

The book is to be opened in connection with Divine Judgment; (Dan7:10) (Rev20:12) In the new Testament, those redeemed by Christ are contained within the book. (Philippines 4) Those not found within the book of life will not enter the kingdom of heaven.

As an interesting corollary, in the ancient world- a person's Name was symbolic of his or hers existence.

According to Sir James George Frazer author of the Golden Bough- one of the most extensive volumes on World mythology, there was such a bond between one's Name and one's Existence "That magic may be wrought on a man just as easily through his name, as through his hair, his nails, or any other material part of his person."

In ancient Egypt to blot out a Name of a record was equivalent to destroying the fact that the person ever existed.

To Quote a Biblical Reference: Jesus had said- "A Good Name is better than a Precious Ointment!"(Ecclesiastes 7:1).

To Quote another Biblical Reference: (Mathew 12:34)

"For I say unto you that for every foolish word which men speak, they will have to answer for it on the day of judgment. For by your words you shall be found guilty'"

During my intense Spiritual Awakening and new found wisdom with names, I found that when the student is ready the teacher will appear.

I found the answers can come within one's self or from other people that might have relevant information to what you are seeking. I discovered in life on Earth there are no random acts! One such Teacher came to the computer store where I was working at. It was my husbands and I own business. This wonderful lady, wonderful author is Shirley Lawrence Blackwell.

She had brought her computer in for repairs one day; and as we got to talking I discovered that she was an author of several books on the power of Words and Numerology.

The books Shirley Lawrence Blackwell had written were exactly what I was looking for. The titles of the books are: Numerology and the English Cabbala; The Secret Science of Numerology the Hidden meaning of Numbers and Letters; and Exploring Numerology Life by the Numbers.

I found reading her books very enlightening due to my being able to look at a name and being as a mirror; I was able to see the acts; the scenes; and the story of the entity or soul carrying the name. The master numbers in her books intrigued me to do the numerology on my own name.

My Total name Vibration JOYCE =22 ANN=11 MORFORD=44.

AS you can see 77 vibration in Numerology is Christ Consciousness. Which means- An intensification of inner wisdom.

THE GIFT

The greatest gift that I could have ever had been given is to walk this Earth in a Christ state of awareness. I can only describe this state as: Joy; Oneness; Peace; and Belonging.

Someone May steal all your earthly belongings in this material Earth;

Which by the way; is only temporary. No one can steal your gift of oneness; your high vibrational energy that you give to the Earth.

If everyone walking the Earth were in this state of oneness of unity we would have "A New Earth" a Utopia; the golden city of light.

In that high vibrational consciousness I am you and you are me. There is no separation, only "Love and Unity! The realization; Love is the only thing that's real on this planet. Higher Consciousness is here now as many begin to awaken to their true self.

I am truly blessed this time around to be able to experience both sides. Being blessed having an EGO and also not having an EGO.

To me your whole part of existence here is find out this truth; that you are much grander than you've ever been told.

The material world with all its trappings; and the false dogma that has kept humanity in a lowly earthly material state is truly a blasphemy in itself.

The Ego can be good provided its use of it is for good and positive intentions. The Human which incarnates here on Earth is born with a false self the Ego personality. It was given a name and will experience the events in its name either it be positive or negative according to its environment growing up. The Ego in its negative aspect is the devil itself. The Ego in its negative state is what causes unhappiness; judgment; hatred; envy; jealousy; greed. It is in this separateness that causes the Ego to compare itself with others which either deems themselves superior or inferior to the self. In all reality we are clearly One and the same; it is unfortunate we were not taught at birth the truth! For if we were taught that all are one! One! One! In the beginning we would have a more peaceful Earth. It is in this negative state of consciousness that the Ego can and will do its most damage to other humans. The true state of your Egoless self is the pure love of God. You are already God hidden behind a false personality and a Name that you took when you came here to Earth.

Humanity is in such ignorance of a person without an Ego; that they look upon you as if you are crazy, or in a parallel universe. Maybe we should be checking in on our brothers and sisters that are in insane asylums now, being miss judged by the Egos who cannot possibly understand an Egoless state. What an oxymoron this is. When someone is full of nothing but love and trying to convey a new truth to the world that can only be achieved in an egoless state; the Ego is in utter disbelief! The Ego can't even conceive of such a thing. The Ego only wants to hold on to its image of being real. The death of the Ego though is exactly what needs to happen in order to bring about a balanced human being. The saying to "die daily" is exactly this.

My point is when an Ego and Egoless person come together it truly is living in two separate realities; One on a lower vibratory state of being, and the other on a higher vibratory state of being.

I am writing this book to make humanity aware and to learn more about this state of being so when one of their family members has it happens to them they will not be in ignorance of what is going on with this person; this soul that is naturally evolving on the Earths plane. The altered high vibrational state of "Christ Consciousness" is what the Master Jesus taught, in this altered state of Oneness with all is the pure love in the heart; the Christ Heart! It starts with the heart, it ends duality and separateness. It ends separateness from God; so you're true God self emerges. Christ manifest's himself within the Human being here and now! Jesus manifested the Christ within himself! It is not a person out there! Christ is in each one of us; he suffers all Human experience inside of us! That is your reason for being here; to find the God within.

Those that look for Christ to return outside of their selves are truly miss-guided. Christ return is within you. As stated in Revelations: Who can open the book and the seven seals to read the book of life. The seven seals are your chakras. The book is the Scroll which is D.N.A.

D.N.A. when you look at it is a ribbon which looks much like a scroll.

Christ reveals himself through you once you have enter a higher vibration; the higher chakras in meditation.

To find the God within; one needs to enter the inner chambers of his/her temple to meditate in silence for your true Christ self to emerge. Meditation allows the higher chakras to open. A lot of Humanity now operates in the lower 3 chakras. To connect with the Christ self the heart chakra must be opened first then the higher centers will open like a flower; in this way you can obtain a connection with your higher self; your Christ self! Jesus state of Christ Consciousness can be accessed with pure desire to experience the state of consciousness that he walked the Earth in as a master of pure love. Your consciousness can be over shadowed by the Christ.

It was in this state of consciousness awareness that Jesus had attained that allowed all his psychic centers to be opened. Jesus was the greatest psychic of all time; for in this state of consciousness all other consciousness is one and can be access for viewing.

For Humans are also Spirit in form nothing can remain secret!

My gift of Nomancy allowed me to see; to feel; to hear the acts; the scenes; and the story by their Names. By reading of the Name of an entity I can feel their emotions, whether it be for good intent or not good intent, I can read the words they will say, the things that they do, my Consciousness is one with their Consciousness. I mirror their Consciousness in writing.

Consciousness does not die. D.N.A. does not die. The body may die but Consciousness lives forever. In a state of oneness, you may read any consciousness because there is no separation.

My Gift in this lifetime is to be able to read the names of others. My gift is to be able to access their consciousness on the other side of the veil. I am a medium to the other side. I am using multidimensional D.N.A. to obtain information in a non-linear way. I am able to communicate with multidimensional entities in a quantum state. In a metaphysical community it would be seen as life is a hologram within this 3-D illusion. We all have an act, a scene, a story to play! My gift of awakening from the 3-D holographic illusion to know what I truly am. I am a light being having a human experience in a physical body with limitations. My awakening enabled me to see and know that my D.N.A. is not of this planet. I am classified as a wanderer here on Earth. My gift was to uplift humanities vibration of their selves.

This is the message; CHANGE YOUR NAME CHANGE YOUR LIFE!

TO THE READERS AWARENESS

The Act, The Scenes, The Stories, that you are about to read in this book; are not to be read in the literal way.

Although you are reading the words on the page you want to READ it as an OBJECTIVE OBSERVER.

As an objective observer you will view it as an audience would in a Movie Theater; watching the ACTS, the SCENES, the STORY; The STORY LINE.

For as Shakespeare once said: Quote "Everyone's an actor and all the world is a stage."

The Reader must be Aware that words are not just merely words. When you read them it invokes a feeling, a sound, it invokes a picture in the mind of the reader of the situation going on in an event. The spoken word emits a vibration of its own.

I want to instill to the reader that my Gift of Nomancy is a rare form of divination. Nomancy means: The art or practice of divining the destiny of persons by the letters which form their name. It is a form of clairvoyance that allows me to see, and feel, and hear the acts, the scenes, and the story in your name. Your name is an energy that can be picked up on by some who is a Sensitive to energy. It allows me to be a mirror to that person's (soul) life event. Any word emits an energy and vibration. It can also be a fictitious name as well. I will be showing real life events using a fictitious name.

I want to instill in the reader that the feelings that surround the act, the scenes, the story are not my own. I am projecting the Soul who carries that name vibration and any of the other souls involved. I am projecting their feelings and emotions in the reading. I am very empathic and can pick up emotions of everyone around me in everyday life. In having this gift of Empathy it allows me to pick up on the person (souls) feelings and projecting it on to the paper in the Acts, the Scenes, and the Story; by only using the letters of tier name.

This Rare art of Nomancy that I possess allows my D.N.A. to unravel the letters of the name to show me the Acts, The Scenes, The Story.

When divining a person, place, or thing using Nomancy, you can only use the letters that are present in the name. If you only have for example one A in your name you can only use it once in any given word. If you have two "T" s in your name, then you can use it two times in one word etc.

What I discovered was a Name seems to have a macro code in it that is not detectible by the average eye by looking at it. It seems that a Name has a story within its self that is not a random act. In fact, it is your life unfolding within your own name. Your name creates your reality here on Earth. I found your name entangled with someone else's name will give you the acts, the scenes, and the story with that person.

The Art or Science that I do is channeled works of Art or Science. This is my Passion and the quest of why or how and the reasoning behind it begins in the next chapter.

SECTION 2

REVEALING THE TECHNIQUE OF NOMANCY

Here in this chapter you will find my channeled name arts or science by only using the letters of the name. The term Channeling; meaning information coming from the inside of myself; my D.N.A. the other side of the veil. Communication is from within from using quantum= Multidimensional sources within the D.N.A.

Each name art is true to itself. The acts, the scenes, the story are in the name as a macro code. This is my gift unscrambling the story intuitively. By using my innate; my intuition.

THE NAME ARTS
OR
SCIENCE OF THE SPOKEN WORD EXAMPLES

OPHIDIAN CURRENTS

C=(K)SOUND

PH=(F)SOUND

READ A RARE ART, A TRUE ART OF ONE!

SHE CAN C A SECRET IN THIS ART!

I READ AN ART IN OPHIDIAN CURRENTS. I RAISE THE OPHIDIAN CURRENTS IN I TO CREATE THIS ART! SOPHIA (Wisdom) RISES IN THE CURRENTS. SEE THE NORTH POLE STAR.A NORTH POLE STAR A LINES IN AN EARTH POLE STAR INSIDE U! RAISE THE SOUTH CURRENT UP TO THE NORTH CURRENT. U R A STAR!

NO LIES IN THIS ART.THIS ART IS ONE, THIS ART IS A, HE/SHE ART. I; A SACRED CURRENT

AN ANCIENT CURRENT IN I. A SACRED ANCIENT EUCHARIST! = (A SACREMENT TO CHRIST, TO COMMUNION OF BREAD AND WINE.) ONE IS TO CHANT AN ANTIPHON= (A PSALMS VERSE, A SONG) IN UNION OF THE ONE HEART! I RAISE THE SPIRIT UP! UP!

I, A RED SERPENT, I A STAR PH-IRE! PHI (THE DIVINE RATIO) A CURRENT NEAR THE ANUS! I HADIT AND NUIT ARE ONE! THE EARTH AND I ARE ONE! IN THE OPHIDIAN CURRENT, HIDES NECTAR. A SHE AND HE CAN RAISE THE OPHIDIAN CURRENTS IN COITUS. COITUS IS A SACRED ART!

D.N.A HIDES IN OPHIDIAN CURRENTS. PI-HIDES IN D.N.A., PI IS A CODE IN D.N.A. C? D.N.A. IS A HEART TUNER. HEAR THE TONES, HEAR THE NOTES, HEAR THE CHORDS. HEAR THE HARP.

HORUS=CHRIST IN D.N.A., IN OPHIDIAN CURRENTS. I CHRIST IS IN D.N.A.!

RAISE CHRIST IN D.N.A! U CAN HEAR NATURE SING! STAND UP! IT'S TRUE!

RAISE THE CURRENTS IN THE SPINE TO CURE D.N.A. RAISE THE OPHIDIAN CURRENTS IN THE SPINE-IT RE-SIDES IN THE SPINE. SEND THE SERPENT CURRENTS UP NORTH TO THE HEART, TO THE HEAD.CARRIE THE CURRENTS UP THE HEART TO THE HEAD, THEN ARC IT! ANC(k)H IT! THE OPHIDIAN CURRENTS CAN RE-CODE D.N.A. IT RE-DOES D.N.A.

CHRIST SPIRIT IS IN D.N.A A HEART; AN EAR NEAR CHRIST CAN CURE. TO RAISE A CURRENT IN D.N.A. OPHIDIAN CURRENTS DO NOT RISE IN A HEART THAT HATES! A PURE HEART CAN RISE TO CHRIST! TO CURE A HEART RAISE CHRIST IN D.N.A.! A HEART IS SACRED. RAISE A

PHI-RE IN A HEART; RAISE THE PHI-RE TO THE HEAD, A SHE/HE READS THIS IN TRANCE!

A SHE/HE CAN RAISE THE OPHIDIAN CURRENTS IN A TRANCE. (MEDITATION)

U, ARE A HE/SHE, AN ORPHANED = (DEPRIVED OF PARENTS BY DEATH OR DISERTION) HERE IS A SHE/HE, SHE CAN READ D.N.A. CAN READ THIS ART.

D.N.A. DOES NOT DIE! NO DEATH IN D.N.A. NO DEATH IN THE OPHIDIAN CURRENTS.

DO NOT REACT TO HATE. NO REACTIONS TO HATE TO RAISE D.N.A. TO RAISE A SACRED CURRENT! D.N.A, R.N.A A RADIO TURNED ON!

OPHIDIAN CURRENTS RISE IN URANUS, A STAR RISES IN URANUS! (AQUARIUS) URANUS IS A SUN STAR-CHRIST IS A SON (OF) THE SUN.TRUE! THIS IS SO!

HORUS=CHRIST=SON (OF) A SUN.

U CAN TURN THE TIDES OR CURRENTS, U ARE A SHE/HE. D.N. A= IS/ON = AN OPEN CURRENT=ONE CURRENT.

A CHANNELLED WORK OF ART

OPHIDIAN CURRENTS SYNOPSIS

The Ophidian Currents is what is known as Kundalini energy. Kundalini energy sits near the anus and is coiled 3 times. In mediation you can raise this current up the spine to the heart and Arc it to the head which creates a magnetic field of energy around you. It dissolves the Ego, the small" i" in you.

A pure Heart can raise this current by chanting, by toning, even by reading sacred verses. This current can be raised in seeking the true self with a pure heart wanting to know the truth. When the student is ready the teacher will appear. The Teacher is within your own self! Your Christ Self is there to be awakened. In raising the Kundalini energy within yourself you put on the Christ Consciousness, and you walk the Earth as the Masters have done in the past. When Male and Female unite together in Tantra; which is sacred sex with a pure Heart; kundalini energy rises and creates a third energy. Mediation shuts down the realm of thought; the thinking mind.

D.N.A is ignited and your core self is revealed. Your core self is the Christ Consciousness. You are at-one- ment with all and everything.

We all have Male and female energy currents inside us, they are called Ida=Feminine energy and Pingala= Masculine energy which when kundalini rises it joins these currents into one current.

When one raises the Christ Consciousness there is no separation, Health is restored, All is Oneness, Peace, Joy and Belonging! This is how we are to create Peace on Earth. In Oneness there is no polarity or duality all is one.

THE SEAL OF SOLOMON THE STAR OF DAVID

OM!

ME A NAME AMEN!

SAVE ALL ON THE EARTH! ONE MIND-ONE HEART!

THERE IS NOT ONE NAME ON EARTH THAT SHALL HIDE FROM ITS SELF; IT'S FALSE SELF! (EGO).

THE DIVINE ONES IN HEAVEN TELL HER NOT TO FEAR! NO FEAR! TELL ALL THE ONES ON EARTH!

A STAR OF ONE ON THE EARTH! ALL ARE ONE!

ALL ARE FROM HEAVEN! ALL D.N.A. ARE FROM HEAVEN! A SEER SEE'S AND HEARS: ALL ARE DIVINE ON EARTH! A SEER SEE'S THE STAR OF RA. SHE HAS AN EAR NEAR RA! A SEER SEE'S THE THREE STARS OF ORION! A SEER SEE'S-START AT THE EAST, THE THREE STARS OF ORION! SEE

THE SEVEN SISTERS ARE SEVEN STARS! SEE THE STAR LINE FROM EARTH! (PLEIADIANS)

A SEER HERE LIFT'S THE VEIL OF A NAME! SHE READS THE FORMS ON EARTH FROM THE LETTER'S THAT FORM THE NAME, SEE?

SEE HER DIVINE NAME ON EARTH; HER D.N.A IS ALIEN TO EARTH.

SHE READS: THERE ARE NO DEMONS IN THE STARS!

THIS SEER DIVINES FROM THE LETTERS IN A NAME! (SHE CHANNELS) SHE HEARS IT ON THE INSIDE OF HER MIND! (Clairvoyant, Clairaudient, Clair cognizant). SHE READS ALL THE NAMES ON EARTH!

SHE HAS AN EAR NEAR HER OVER-SELF. (Over Soul OR Higher Self). THIS SEER HERE IS FROM THE SEVEN SISTERS OF HEAVEN; THE SEVEN ETERNAL STARS! HER D.N.A IS FROM THE SEVEN SISTERS! SHE HEARS THE NOTES AND THE TONES FROM HEAVEN; FROM THE STARS.

SHE HEARS: IT IS SAFE IN THIS LIFE TIME TO REVEAL THIS ART OF NAMES. IN ANOTHER LIFE MEN SET HER ON FIRE IN SALEM MASS SEE? SHE IS STILL HERE; D.N.A NEVER DIES!

THIS SEER HERE FELL FROM THE THIRTEENTH STAR TO EARTH. (MEANS CAME FROM THE THIRTEENTH DIMENSION) THE THIRTEENTH

STAR IS A STAR OF LOVE! SHE RODE THE STARTETRAHEDRONS TO EARTH! SEE?

THIS ART IS FROM THE FATHER OF THE STARS AND MOTHER EARTH! HONOR THE FATHER (MIND) IN HEAVEN AND MOTHER EARTH! (HEART-EMOTIONS)

THE STAR OF DAVID IS SOLOMONS'S FATHER!

IN THIS NAME ART "THE SEAL OF SOLOMON-THE STAR OF DAVID REVEAL'S THE STAR TETRAHEDRON'S! D,N.A. IS A RADIO ON! D.N.A. IS ALTERED FROM THE FATHER'S OF THE HEAVENS!

A SEER SEE'S HEAVEN'S FATHERS SEED'S THE EARTH. THE FATHER'S SEMEN OF THE HEAVENS, OF THE STARS, SEED'S THE FEMALE OVARIES OF THE EARTH. THE FATHER OF THE HEAVENS, THE STAR'S AND THE FEMALES OF THE EARTH MATED. A SEER SEE'S ALIEN D.N.A. IN THE FEMALES; IN THE MOTHER'S OF THE EARTH! ALL ARE ONE IN HEAVEN AND ON EARTH!

THIS ART REVEALS THE ONE-NESS OF THE STARS AND THE EARTH! THE MIND OF THE FORM IS ERASED ON EARTH; THE MIND IS VEILED ON EARTH! SHE READ'S THE MIND IS ONE! THE MIND DIVIDED INTO MALE AND FEMALE; IT IS VIELED!

SATAN IS (EGO) THE FALSE SELF ON EARTH. SATAN (EGO) IS THE MIND DIVIDED! RAISE THE

MIND AND HEART TO ONE-NESS! (HIGHER Consciousness)

STILL THE MIND, SILENT THE MIND, MOVE THE MIND INTO LOVE AND ONE-NESS= MEDITATE ON IT! MEDITATE ON THE STAR TETRAHEDRONS!

A SEER SEE'S THE STAR TETRAHEDRON'S ROTATE; A MALE STAR AND A FEMALE STAR TETRAHEDRON.A SEER SEE'S THERE ARE THREE LEVELS OF MIND! THE THIRD LEVEL OF MIND = ONENESS! (mind = consciousness) MIND IS MALE, EMOTION'S ARE FEMALE. LOVE IS A SOLVENT; IT IS LOVE THAT STEERS THE STAR'S! LOVE ROTATE'S THE STAR TETRAHEDRON'S. THE FASTER IT ROTATES; IT HEALS THE MIND; IT HEALS THE HEART.THE FASTER IT ROTATES ALL ARE ONE! ONENESS! MEDITATION ON THE STAR TERAHEDRON'S MOVE'S IT FASTER! IT MOVES THE MIND INTO ONENESS; IT SOFTENS A HEART THAT HATES INTO THE EMOTION'S OF LOVE.THE STAR TERAHEDRON'S MOVE'S FASTER IN A HEART THAT LOVE'S! IF THERE IS NO LOVE THE STAR TETRAHEDRON'S DIE. SO HEAL A HEART THAT HATES AND RAISE A FIRE! A STAR FIRE! A FORM IS A SHELL FOR THE MIND AND EMOTIONS ON EARTH. EARTH IS A TEST OF LOVE! EARTH IS A TEST OF THE MIND! THIS SEER; SHE HEARS THE NOTE'S AND THE TONES OF THE STAR TETRAHEDRON; OF D.N.A, DO-RE-MI-FA-SO-LA-TI-DO! HEAR THE NOTES AND THE TONES AND

RISE! THESE NOTES AND TONES FORM THE STAR TETRAHEDRON'S~

THIS ART REVEALS D.N.A NEVER DIES!

THIS NAME ART IS FROM THE OTHER SIDE OF THE VIEL IN D.N.A. IN D.N.A! SEE HEAVEN! A SEER SEE, SOONER OR LATER I SEE HEAVEN ON EARTH!

THIS NAME ART IS REAL! NOT A LIE! THERE ARE NO LIES IN THIS NAME ART.THIS NAME ART IS FROM THE DIVINE ONES IN HEAVEN!

SEE ALL THE LOVE FROM HEAVEN IN THIS NAME ART! A SEER HERE AND HER DEAMON= (HIGHER SELF, HOLY GUARDIAN ANGEL) ARE ONE! A SHE/HE A SEER HEARS HER DEAMON!

SEE A STAR, STAR OF SOLOMON, STAR OF DAVID

THE SEALS AND STARS ARE THE SAME. THE NAMES ALL MEAN THE SAME.

I AM A SEER; I SEE THIS IN THIS ART, THIS NAME ART OF THE ONE SON! SOL! THE ONE STAR! A HE/SHE SEE'S THE SAME SEAL, SAME STAR.

THE SAME STAR IS IN HER NAME ART. THE SAME STAR IS IN HER STAR ART! (Astrology chart) SEE HER STAR ART!(ASTROLGY CHART)

THIS STAR IS THE STAR OF HEAVEN; SEE IT SHINE IN THE HEAVENS! SEE IT IN THE HEAVENS! SOLOMON AND DAVID ARE THE SAME. ALL OF THE STARS ARE THE SAME.

A SEER IS HERE, A SHE/HE SEES THIS; A SHE/HE DOES THE NAME ART FOR ALL. SEE THIS IN HER NAME ART! SEE HER STAR ON EARTH! SEE HER EARTH STAR! (ASTROLOGY CHART) READ THIS IN HER NAME ART!

SHE FEELS THE ONE HEART, SHE SEES THE ONE HEART, SHE SEE'S THE DIVINE IN ALL.

THE STAR IS TO REMIND THE EARTH OF HEAVEN! THE STAR IS TO REMIND EARTH THAT THEIR D.N.A IS FROM THE ALIENS OF THE HEAVENS, THE STAR'S!

HEAVEN ON THIS EARTH STARTS WITH THE HEART; RAISE A FIRE TO THE HEART!

SEE THE NAME OF ADAM IN THIS ART, SEE THE NAME OF EVE IN THIS ART, HEAVEN AND EARTH ARE ONE. HEAVEN AND EARTH AN EDEN! HEAVEN AND EARTH ARE EVEN. (Mind and Emotions in balance)

THE DIVINE SEED FROM HEAVEN IS IN ALL, IN ADAM, IN EVE. THE DIVINE SEED IS D.N.A. A NAME IS A FORM OF D.N.A. THE STARS OF HEAVEN SEEDED THE EARTH. THIS SEER SEE-STARMEN FROM HEAVEN SEEDED THE EARTH! STARMEN SEEDED D.N.A.

THIS NAME ART REVEALS THE FALL OF ADAM AND EVE FROM HEAVEN TO EARTH. THE FALL OF D.N.A.

THE DIVINE SEED FROM HEAVEN DI-VIDED INTO MALE AND FEMALE D.N.A. AND FELL TO EARTH.

IN HEAVEN ALL ARE ONE MALE/FEMALE.

ADAM AND EVE FELL TO EARTH TO SEE THE FORMS, TO FEEL THE FORMS, TO HEAR THE FORMS, TO TASTE THE FORMS OF THE MATERIAL EARTH. TO SEE AND FEEL THE TREES, TO SEE AND FEEL THE ANIMALS, TO SEE THE FISH IN THE SEA, TO SEE AND HEAR THE SEA, TO SEE AND FEEL THE LOVE OF THE ONE SON SOL! TO FEEL THE LOVE OF THE HEART.TO FEEL THE HATE OF THE HEART.

THE SEAL OF SOLOMON AND THE STAR OF DAVID, IS TO REMIND ADAM AND EVE OF HEAVEN, ADAM AND EVE- ADAM=MIND-EVE=HEART. THE HEART FELL TO MIND. HEART AND MIND- MALE AND FEMALE LOST SITE OF THIS STAR! THE SEAL, THE STAR IS TO REMIND ADAM AND EVE OF ONE FOOT IN HEAVEN, ONE FOOT ON EARTH, ADAM AND EVE LOST SITE OF THIS IN THE MATERIAL FORMS OF THE EARTH! ONE FOOT IN HEAVEN-ONE FOOT ON EARTH MEANS EVEN!

ADAM AND EVE LOST SITE OF THE ONE-NESS OF THEIR D.N.A, THE DIVINE SEED FROM HEAVEN.

THE FATHER SOL (SUN), THE MOTHER EARTH IS ONE, SEE THIS IN THE TREES, THE LEAVES, THE ANIMALS ALL FORMS OF THIS EARTH ARE ONE IN HEAVEN, ONE ON EARTH!

THE LETTERS THAT MALES AND FEMALES READ, SEE, AND HEAR, FORM THE FORMS OF THE MATERIAL EARTH. THE LETTERS ARE DIVINE, THE LETTERS ARE ALIVE. THE LETTERS FORM THE MATERIAL EARTH. SEE, READ, HEAR THE LETTERS OF LOVE NOT HATE. TO HATE, IS TO DIVIDE, TO LOVE IS TO FEEL, SEE, AND HEAR THE ONE-NESS OF ALL! IT STARTS IN THE HEART.

A SEER SEE THE TREE OF LIFE SAME AS THE TAROT.

READ THE TAROT, IT REVEALS ALL TO THE EARTH. IN THE TAROT, READ THE FOOL, (THAT IS U). THE FOOL AND THE ROAD HE/SHE TRAVELS.

HEAVEN IS INSIDE A MALE, INSIDE A FEMALE,

THE EARTH HAS FAILED TO SEE THAT THE LETTERS FORM THE EARTH; LETTERS FORM THE HEAVENS. LETTERS HIDE IN THE FORM, LETTERS REVEALS THE FORM TO ITS SELF, TO ITS ELF! LETTERS REVEAL THE LOVE AND THE HATE OF ITS SELF, OF ITS ELF. LETTERS REVEAL THE D.N.A. OF ITSELF, ITS ELF! LETTERS REVEALS A FORM, ALL FORMS! THE TAROT REVEALS D.N.A.!

THERE IS NOT ONE FORM, NOT ONE NAME THAT SHALL HIDE IN ITS SELF THAT SHALL HIDE FROM ITS ELF. ALL IS REVEALED IN A NAME! ALL IS REVEALED IN THE D.N.A OF A NAME! REVEALED IN A NAME OF A FORM, REVEALED IN THE NAME OF THE ANIMALS, REVEALED IN A NAME OF A

MALE AND A FEMALE.SEE? ALL IS REVEALED TO HEAVEN FROM THE NAME. LOVE AND HATE IS REVEALED IN A NAME. NOT ONE NAME SHALL HIDE FROM HEAVEN. HEAVEN SEES ALL! HEAVEN HEARS ALL. HEAVEN AND HELL IS INSIDE THE NAME OF ITS SELF, IT'S ELF! A SEER, A SHE, A SHE/HE REVEALS THIS IN THIS NAME ART! SEE IT. READ IT. SHE FEELS THE LOVE OF ONE, OF ALL.

THERE IS LOVE IN A NAME, THERE IS HATE IN A NAME, THERE IS EVIL AND DEMON IN A NAME, THERE IS DIVINE IN A NAME.

D.N.A. THE TREE OF LIFE, SEE THE ART OF THE TREE OF LIFE, SEE THE ART OF THE TAROT, THE FOOL THAT TRAVELS THE EARTH, THE FOOL THAT LOST SITE OF THE SEAL AND THE STAR. THE FOOL, THAT HAS LOST SITE OF HEAVEN ON EARTH. THE FOOL THAT SEES THE EARTH AS DIVIDED.THIS NAME ART REVEALES A LOT!

THIS NAME ART REVEALS THE ART OF A NAME, THE ART OF D.N.A. A NAME, AN ART, AMEN!

THERE IS MORE TO REVEAL IN THIS NAME ART.

A DEAR ONE READS OF THE SEVEN SEALS IN REVELATIONS, THE SEALS OF MALE, FEMALE IS HIS/HERS D.N.A. STRANDS. D.N.A HAVE SEVERAL TONES, HEAR THE TONES, FEEL THE TONES, RELEASE THE SEALS OF D.N.A. HEAR THESE TONES, DO, RE, ME, FA, SO, LA, TE, DO, D.N.A IS THE DIVINE SEED! RELEASE THE SEALS, LOVE; EVEN LOVE THE ONES THAT HATE! RISE TO

HEAVEN LIKE THE ROSE, OR FALL TO EARTH LIKE THE LEAVES ON A TREE IN THE FALL.THE TEST OF LIFE, IS A TEST OF LOVE, THE ONE HEART! IN THIS NAME ART OF HEAVEN AND EARTH; SHE READS OF THE SEMEN AND OVARIES IN THIS NAME ART. SOL= (THE SUN) =MALE. MOON= (MOTHER) FEMALE, AS ONE IN LOVE, FORM THE FORMS.

THIS ART REVEALS THE FORE- FATHERS ON THE EARTH. THE FREEMASONS HIDE THIS FROM THE EARTH. THE FREEMASONS HIDE THIS FROM ALL ON EARTH! THIS ART REVEALS THE SEVEN HILLS OF ROME. SEE? ALL IS THE SAME IN THE STARS AS IT IS ON EARTH!

SEE? A LOT IS HIDDEN IN A NAME ART.A NAME A VEIL. THE LETTERS OF A NAME MADE THE FORMS REAL ON THE EARTH!

SEE A NAME ART, READ A NAME ART, AND LIFT THE VEIL OF THE FORM AMEN! ME A NAME AMEN!

SEND LOVE TO THE ONE'S THAT HATE!

SEE A MASTER'S STAR ON EARTH= ELEVEN TIMES SEVEN= 77. (CHRIST CONSCIOUSNESS)

A MASTER'S STAR IS THE STAR TETRAHEDRONS!

TRAIN THE MIND AND THE HEART TO LOVE ALL!

AUTHOR: JOYCE ANN MORFORD

22 11 44 = 77

A CHANNELED NAME ART

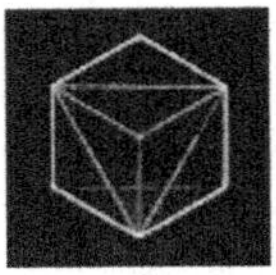

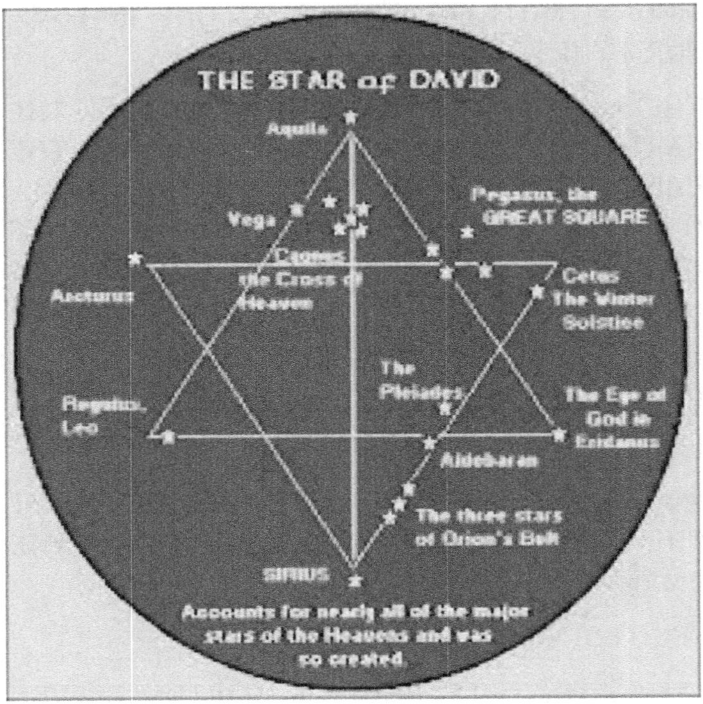

THE SEAL OF SOLOMON THE STAR OF DAVID SYNOPSIS

This channel of this Name Art explains the real nature of the Seal of Solomon- the Star of David.

It is about our D.N.A. and where it originates from.

Our D.N.A was altered by alien beings in the Pleiades. They are the ones that seeded our DN.A. They are the ones who gave us our spiritual attributes.

The three stars of Orion is the guide post in the heavens that leads to the Christ Star it is called the Star of Ra.

The Star Tetrahedrons that rotate around your body; this energy field is your light body. The emotions of love rotate the Star Tetrahedrons faster. The faster it rotates the Male and Female parts of the brain are joined as one. The female side is your intuition or Innate mind your subconscious mind; the Male is is the logical side that many use today in Duality Consciousness. Many do not use their Female side of the brain at all. The intuitive side, the female side knows the truth of everything. When activated the female side informs the male side of the actual truth of everything.

There is a macro code in a name that describes the acts, the scenes, and the story of each and every name.

When the star tetrahedrons are spinning fast enough it elevates your consciousness to higher state of being. This Consciousness is called Christ Consciousness. Christ Consciousness is your mind elevated to a quantum state while still in the physical body.

Jesus manifested this state of consciousness in him. And likewise he told us we could do the same. This is true. Our religions of today made Jesus a divine being that no human could attain to. This is the big lie; there big secret they have kept from all humanity. These are crimes against humanities birth right! This is an abomination of females for centuries! They wanted control over the people; and they wanted Men to rule; when in actuality we are equal as this name art portrays.

The religious leaders of the Earth are teaching people that the second coming of Christ is Jesus return. I will quote Jesus here "I have to go now so the comforter can come within you "The second coming of Christ is manifested in you! I know because I have been in this unity state of consciousness since the year 1998.

The name art reveals that consciousness is in a form. The reason your consciousness resides in a form is to experience the five senses. They are, to see, to taste, to touch, to hear, to smell.

You travel the tree of life to get here to experience this state of duality and polarity.

In reality you are androgynous; when you come to earth you either come in as male or female or even transgender.

Your purpose here is to find the truth within you.

This name art reveals your true origins, yes you were seeded by an alien race and yes all are ONE! There was never supposed to be an "Elite" group of people! There was never supposed to be a difference in equality between male and female. Women suffered much in a male powered society!

This name art reveals the sacredness of the male and female sexual organs, Male seeds the Female, the female births the form that the soul will house in. You will See this in an up and coming name of "The Holy Grail"

Letters form your reality here, First there is the Word and the Breath= Sound! My name arts show the entanglement of relationships with one another whether it be for Karma or Dharma. Letters- Lets a Seer See.

Amen= me a name Amen -and so it is on Earth!

THE NAME ARTS OR SCIENCE OF A PLACE EXAMPLES

RENNES LE CHATEAU

C=(K) SOUND

LET US SHARE A CHAT!

LET US SHARE A SECRET!

LET US SHARE A SECRET CHAT!

LET US SHARE A SECRET ART!

LET US SEE THE SECRET CHART THERE!

A SEER CAN SEE THE CATHARS THERE!

A SEER HERE CAN SEE AND HEAR A SECRET RACE. A SEER CHANNELS A SECRET RACE AT THE EARTHS CENTER. SHE CHANNELS US THE ANSHAR!

A SEER HERE SHARE A TRUE ART!

SHE A SEER CAN HEAL' SHE'S A HEALER!

A SHE/HE CAN CHANNEL AN ART. A SHE/HE CHANNEL'S A TEACHER. RA AND THE ANSHAR THE TEACHER'S SEE?

A SEER CAN SEE AND HEAR THE ANSHAR. THE ANSHAR ARE THE TEACHERS! SEE THE ANSHARS

SATURN SEAL! SEE SATURN! THE ANSHAR HAS SATURN AS A SEAL (I see an amulet on their chest) CHEST! I A SEER SEE THREE RACES AT THE EARTH CENTER. I SEE A STAR SEAL! SEE A STAR SEAL (on their chest) SEE THE TUNNELS; A SHE/HE REST AT THE EARTH'S CENTER! (inner Earth) ENTER AT THE EAST!

AN EAR NEAR THE HEART; AN EAR NEAR NATURE.A SEER SEE THE TRUE NATURE HERE AT RENNES LE CHATEAU!

A SEER HERE CAN CHANNEL RA! CAN CHANNEL THE ANSHAR! (Ra a race near Saturn) RA NEAR SATURN! RA A SUN STAR SEE? SEE RA! (In the law of one RA is a social memory complex that is located in the rings of Saturn.8th dimension. A social memory complex is a group soul). SEE A STAR ART! SEE HER STAR CHART! A SEER CAN SEE A STAR

CHART AT RENNES LE CHATEAU! (SEE A STAR MAP)

A SEER CAN HEAR US THE TEACHER'S AT THE EARTH'S CENTER.

A SEER CAN SEE U.S.A HAS TUNNELS. U.S.A. HAS SECRET TUNNELS! THE ANSHARS HAS TUNNEL'S AT U.S.A.

SEE THE CLUE'S? SEE NATURE'S TREE'S? TREE'S ARE A NATURAL SHELTER.

A SEER CAN SEE THE NUN'S AT RENNES LE CHATEAU. THE NUNS SHARE A SECRET.

A CHATEAU, A CASTLE, NEAR RA, = (THE HIDDEN SUN) THE STAR THERE A SHE/HE, HE/SHE, CAN SEE A SECRET STAR SEAL. (The star is the pentagram of the layout of the churches that you can see from above on a map) SEE HER STARS THERE. SEE THE EARTH STAR ART AT RENNES LE CHATEAU! SEE THE STAR, THE EARTHS CENTER! NATURE IS A SHE, NATURE IS THE ART, U CAN SEE HER! U ARE HER!

AT RENNES LE CHATEAU A TEACHER, TEACH.

THE TEACHER, A SHE, THE TEACHER A HE.

U SEE THE TEACHERS, THE TEACHERS SEE U! U A TEACHER CAN TEACH.

THE TEACHERS CAN HEAL U

A TEACHERS HEART NEAR U, A HE, A SHE

AT RENNES LE CHATEAU, A CHART ART CAN HEAL; NATURES ART CAN HEAL. SHE= NATURE CAN HEAL!

THE CHART AT RENNES LE CHATEAU, A SECRET SEAL, A SECRET STAR SEAL, ASECRET SATURN SEAL AN ET'S SEAL.THE ET'S CAN HEAL U THERE. A SEER SEE THE SECRET TUNNELS AT RENNES LE CHATEAU! SEE THE SECRET CANALS!

HE LET A SEER SEE, SHE LET A SEER SEE,

THE TEACHERS, LET U, A SEER SEE.

U ARE A SHE/HE. U CAN HEAL THE EARTH

THE TEACHERS CAN HEAL THE EARTH, A CHART TO SEE AT RENNES LE CHATEAU.

HEAR HER HEART; SEE HER ART AT RENNES LE CHATEAU.SHE HAS A CLEAN HEART;SHA HAS A SERENE HEART.

A SERENE HEART CAN HEAL U, A CLEAN HEART CAN HEAL U

A SECRET CHART AT RENNES LE CHATEAU, U SEE A SECRET CREST.A CREST ON A CHART.A CREST AT RENNE LE CHATEAU.SEE AN ALTAR THERE.

SEE THE SECRET CHART, A TEACHERS CHART. U SEE A SECRET CHEST AND ARC!

A SECRET ARC, A SECRET ART CHART AT RENNES LE CHATEAU

LEARN A SECRET AT RENNES LE CHATEAU, THE CHART, A STAR CHART, LEARN AN SEE THE STAR CHART,

AT THE EARTH CENTER; A STAR ARC (merkaba) A TSAR (KING) THERE. A TSAR HAS A SECRET CHART THERE! SEE A TSAR SECRET CHEST! A SEER SEE THE CATHARS TREASURE THERE AT RENNES LE CHATEAU!

SEE THE STAR AT RENNES LE CHATEAU, = (PENTAGRAM ON THE EARTH FROM THE CHURCHES) A CLEAR STAR THERE.

RENNES LE CHATEAU, A CASTLE THAT HAS A SECRE.SATURN; THE CENTRAL SUN, SATURN A TSAR (KING)

REACH RA, REACH SATURN, AT SUN/SET REACH A STAR AT THE EARTHS CENTER, CAST A NET NEAR THE CENTER, THE CENTER A STAR. U ARE A STAR.

ENTER THE STAR ARC (merkaba) AT THE EAST, AT THE EAST A HEART. ENTER THE TUNNELS, THE CANAL AT THE EAST.

A SUN THERE, SATURN THERE. A HEART CENTER, THE ANSHAR TEACH U THE STAR ART,

SHE ENTER'S A TRANCE, SHE ENTER'S A STAR(merkaba)(to) US! ENTER AT THE EAST. REST AT THE CENTER= (BALANCE) EARTH. THE ANSHAR HAS A SUN AT THE EARTH'S CENTER.

HEAL A HEART THAT HATES, A SATAN HEART-A SERENE HEART HEAL A HEART THAT HATES!

CHARLES, (CHARLES PERRAULT WAS A 17TH CENTURY POET AND WRITER HE WROTE SLEEPING BEAUTY) HE SEE A SECRET ARC, A SECRET CHART, A SECRET CHEST, AT THE EARTHS CENTER. HE A SEER, HE ENTER A TRANCE, SEE THE SECRET ARC, THE SECRET CHART, THE SECRET CHEST.U ARE A CHARLES, HE SHARE A SECRET.CHARLES A HE/SHE.

LET THE EARS HEAR A TUNE.

SATURN A SUN STAR, SEE THE SEAL (SATURN IS ASSOCIATED WITH THE SEAL OF SOLOMON, 6 POINTED STAR)

RENNES LE CHATEAU; SEE THE REAL EARTHS CENTER.

UR; A HE/SHE A SEER. U SHARE A SEAL, A SATURN SEAL, A SUN SEAL, U SHARE THE ART, THE SUN ART AT RENNES LE CHATEAU! SHE/HE CAN SEE US AT THE EARTH CENTER!

RENNES LE CHATEAU SYNOPSIS

I want to say that even a name of a place has events associated with it. In fact this is the first Name art that I had ever channeled. By only using the letters of the name I began to know about this place, as if in a conversation. Letters lets a Seer see is in fact the truth.

The conversation that came out in this name art was of a social memory complex named "RA" (a group soul that is near Saturn in the 8th dimension). The name itself tells a story about a secret there. A secret chest, it tells of an alter there that had a chart enclosed in it; and also the tunnels around Rennes Le Chateau, and of a star. The pentagram is outline in the landscape that connects the churches. It talks about a tsar (a king there) Someone of royalty. It talks about a she (It could be Mary Magdalene herself that was buried there). In another channeled art I did was Francois Berenger Sauniere who is a priest connected with

Rennes Le Chateau; which I will include here in this name art section.

It tells of a way to raise consciousness. A heart that hates is a low vibrational life form. By healing a heart of hate, you purify and heal the heart. A pure heart a grail heart can heal you, which means raising your consciousness to the state of oneness. Ra says, one, one one, (which means we are all one, and in that high vibrational consciousness there is no separation you become the "Christ Consciousness" To me that is the holy grail, to seek a higher state of consciousness. The statement She/he is a state of balance within, which creates the Christed State of consciousness. You are born anew.

In the name at of Rennes, it tells of a Castle that is there called (Magdala Tower). It tells of the tunnels and the Cathar's treasure under Rennes Le Chateau.

So a name is true to itself. Nothing can hide from the name of itself.

THE UNITED STATES OF AMERICA

C= (K) SOUND

OM!

AMEN!

I SEE THE TRUTH HIDE'S IN A NAME!

A SEER READ'S IN A TRANCE STATE OF MIND.

ME A NAME AMEN! CHRIST IS HERE ON EARTH! CHRIST IS MANIFESTED ON EARTH IN THE HUMAN FORM; IN D.N.A. D.N.A. RISE'S, = (means increasing) D.N.A. IS INCREASED. D.N.A. IS INCREASED FROM THE RADIUM AND RADIATION FROM THE CENTER OF THE SUN, FROM THE CREATOR. THE CENTER OF THE SUN HIT HUMAN'S DNA AND UN-LOC(K)S IT. I SEE D.N.A. IN HUMAN'S IS INCREASED! I SEE THEHUMANS ON THE EARTH ASCEND AS ONE; SEE ASCENSION! THE HUMANS ON EARTH ARE FROM THE CREATOR; THE HUMANS ON EARTH ARE CO-CREATORS! THE HUMANS ON EARTH ARE ONE! D.N.A. DOES NOT DIE!

READ A RARE ART FORM. I CAN SEE (Clairvoyance) I CAN HEAR (Clairaudience) AN OMEN IN A NAME. A MIND IN CHRIST- IN ONE-NESS CAN READ THE NAMES! NO ONE CAN HIDE FROM THEIR NAME! I CAN SEE THE ACTS; THE SCENES; THE STORIES ARE IN THE NAME.

I SEE AND HEAR-NO FEAR! NO FEAR OF ROME!

I SEE AND HEAR HUMAN'S REINCARNATE ON THE EARTH!

OUR ANCIENT ANCESTORS OF ROME REINCARANTE IN AMERICA! NO FEAR OF ROME! OUR ROMAN ANCESTORS ARE DEFEATED IN THE UNITED STATES OF AMERICA!

THIS SEER HERE CAN READ THE NAMES ON EARTH.

THE TRUTH IS IN THE NAME! A SEER CAN SEE A MACRO CODE IN THE NAMES ON EARTH.

A SEER CAN SEE THERE ARE NO RANDOM ACTS ON EARTH!

SHE DOES THE NAME ARTS ON EARTH.

A SEER CAN SEE THERE ARE THREE TEIRS (levels) OF MIND ON THE EARTH. SEE: TRANS MIND= (transpersonal) SEE SATANIC MIND= (dual mind) SEE CHRIST MIND= (One-ness).

A CHRISTED MIND = (no ego)

D.N.A IS A RADIO ON!

A SEER CAN SEE "UFO'S ON EARTH! UFO'S INSIDE THE EARTH!

A SEER CAN SEE AND HEAR ET'S INSIDE THE EARTH. I SEE AND HEAR "THE ANSHAR" I SEE AND HEAR "THE STAR ET'S. I SEE "THE DRACONIANS" I SEE MORE THAN ONE RACE OF ET'S INSIDE THE EARTH! I SEE AN ENTRANCE IN ANTARTICA!

THE STARMEN E.T.'S SEEDED D.N.A IN THE HUMAN FORMS ON THE EARTH. (ALIEN'S D.N.A.) I SEE ET'S IN HUMAN FORM ON THE EARTH. NO FEAR! ET'S ARE HERE TO DEFEAT THE "FOURTH REICH" A SEER CAN SEE THE SAME D.N.A. IN HUMANS. A SEER CAN SEE ONE-NESS OF D.N.A.

A NAME IS THE SAME AS D.N.A. A NAME IS STORED IN D.N.A. A SEER CAN SEE AC(K)ASHIC RECORDS ARE STORED IN D.N.A.

A SEER CAN SEE A NAME IS A (life) TIME RECORD STORED IN D.N.A. A NAME CREATES THE ACTS; THE SCENES; THE STORIES OF ONES (life)TIME ON EARTH! HER MIND REACHES THE MONAD THE SOURCE! SHE IS ONE IN THE SOURCE MIND; A MIND OF ONE! THIS SEER HERE RAISES HER D.N.A. TO CHRIST! A SEER SEE: IN CHRIST I CAN SEE "NO MORE MEAN MEN ON EARTH! A SHE/HE IS ONE ON THE EARTH!

SHE RAISES CHRIST IN HER ON EARTH. THE MIND OF CHRIST IS INSIDE U! SHE USES HER STAR TETRAHEDRONS TO RAISE HER MIND

(CONSCIOUSNESS) TO CHRIST. THE EMOTIONS TURNS THE STAR TETRAHEDRONS FASTER. (EMOTIONS OF LOVE NOT FEAR OR HATE)

CHRIST IS A MIND OF ONE. A CHRISTED MIND IS A MIND OF ONE-NESS. THERE IS NO DISEASE IN A CHRISTED MIND. AMERICANS NEED TO UNITE AS ONE IN A CRISIS TO DEFEAT THEM.

A SATANIC MIND (DUAL MIND) IS IN DIS-EASE.

THE FREEMASON'S; THE MASONIC'S DISTORTS THE TRUTH ON EARTH, TO THE HUMAN'S ON EARTH! HEAR THIS! SEE THIS! IT IS THE TRUTH!

THE ANCIENT ANCESTORS OF ROME NEED TO (heal)A HEART THAT HATES!

A HEART THAT HATES CAN-NOT ASCEND TO CHRIST.

A SEER CAN SEE THE TRUTH IN THE NAME. A SEER CAN SEE THAT A SATANIC MIND (is a dual mind a luciferferian mind). A SATANIC MIND IS THE ANTI-CHRIST! I SEE THE ANTI-CHRIST HATES CHRISTIANS! THE ANTI-CHRIST IS RACIST!

A SEER CAN SEE CRIME SCENES IN THE UNITED STATES OF AMERICA! SEE? D.N.A IS A RADIO TURNED ON!

THE ANCIENT ANCESTORS OF ROME ARE IN AMERICA. I CAN SEE OUR FORE-FATHERS ARE THE ONES!

I CAN SEE THEIR NAME-IT IS "THE FOURTH REICH"= (the new world order). I CAN SEE THE "MAFIA" AND THEIR CRIMES TO HUMANS ON EARTH. I CAN SEE THEIR MURDERS OF THOSE THAT DO NOT CONFORM TO THE "FOURTH REICH!

A CHRISTED MIND CAN SEE IT IN ONE-NESS!

A SEER CAN SEE THE FREEMASON'S AND THEIR CRIMES IN AMERICA.THE HEAD FREEMASON'S ARE A FRONT FOR MORE CRIMES IN AMERICA. SEE THEIR DECEIT (LIES); I CAN SEE THEIR MURDERS ON EARTH. I SEE THE HEAD FREEMASON'S OF THE EASTERN STAR!

THE "FOURTH RIECH" ARE HEADS OF AMERICA; ARE HEADS OF STATES, ARE HEADS OF COURT'S; ARE HEADS OF CITIES, ARE HEADS OF FINANCES (BANKS) ARE HEADS OF THE MEDIA ;HEADS OF RADIO; HEAD'S OF C.E.O.'S OF AMERICA; ARE HEADS OF ACTOR'S AND ACTRESS (HOLLYWOOD) ACTORS AND ACTRESS DO IT FOR FAME AND FORTUNE! "THE FOURTH RIECH"ARE HEADS OF E.U. (EUROPE) HEADS OF FRANCE; ARE HEADS OF THE U.N.

A SEER CAN SEE A SECRET HIDES FROM AMERICANS.

A SEER CAN SEE "THE FOURTH RIECH" FUNDS ISIS! I SEE ISIS STRIC(K)ES FRANCE!

A SEER CAN SEE THE STATUE FROM FRANCE IS ISIS. (STATUE OF LIBERTY)ISIS IS HERE IN THE UNITED STATES OF AMERICA. A SEER CAN SEE "ISIS" STRIC(K)ES THE UNITED STATES OF AMERICA!

I CAN SEE TAN MEN IN THE DESSERT AS ISIS. THE U.S TRADES FIREARMS TO ISIS AND OTHERS OUT OF AMERICA. I SEE A DICTATOR FROM A DESSERT RISES.

I SEE A DICTATOR RISES, I SEE FASCISM IN AMERICA.

A SEER CAN SEE OUR FORE-FATHERS THE FREEMASON'S "THE FOURTH RIECH" CREATED THE C.I.A. THE I.R.S. THE U.N. SEE N.A.S.A. SEE N.A.T.O. SEE FEMA. THE "FOURTH RIECH" RUN'S THEM- SEE THE MASTER MIND'S OF THE HUMAN RACE!

THE HEAD FREEMASON'S OF AMERICA TAC(K)E TENDER (MONEY) FROM AMERICANS. SEE THE REASON FOR THE I.R.S. THE I.R.S. TAC(K)ES TENDER(MONEY) FROM AMERICANS TO FUND NASA; TO FUND THE E.T. RACE!

I SEE ATTAC(K)S ON AMERICA FROM THE AIR FROM AIR CRAFT.

I CAN SEE THE MAFIA MURDERS THOSE THAT DO NOT CONFORM TO THE "FOURTH RIECH".

IS THE UNITED STATES OF AMERICA FREE? NO! IS THERE FREEDOM IN AMERICA? NO!

I SEE CRIMES AND SECRETS IN THIS NAME ART.

A SEER CAN SEE THAT THE HEADS (LEADERS) OF THE EARTH FARM KID'S (CHILDREN) FOR COITUS. THE HEADS OF STATES USE KID'S TO FORNICATE! THE "FOURTH RIECH TRAINS KID'S (CHILDREN) UNDER DURESS, TORTURE KID'S UNDER DURESS TO CONFORM TO THEIR ORDER! THE KID'S ARE TRAINED UNDER DURESS TO HATE OTHERS! "THE FOURTH REICH" TORTURE'S THEIR KIDS TO NOT TO TELL NO ONE!

I CAN SEE "THE FOURTH REICH USES S.R.A. (SATANIC RITUAL ABUSE) ON KIDS OF THE "THE FOURTH REICH"! I CAN SEE

THE "FOURTH RIECH" USE'S FEAR TO CONFORM THE HUMANS OF THE EARTH.

I SEE MORE CRIMES: THE HEADS OF AMERICA FORCE INSURANCE ON AMERICANS! INSURANCE IS A SCAM TO TAC(K)E TENDER FROM AMERICANS. I SEE THE MEDIA FORCE THEIR AD'S FOR MEDICATIONS! I SEE AMERICANS ARE FORCED INTO THE H.C. = (HEALTH CARE) REFORM ACT! SEE?

I SEE NO MORE H.C.(health care) CURES FOR AMERICANS!

FORCE INSURANCE ON AMERICANS. IN THE MIND OF CHRIST THERE IS NO DISEASE SO THERE IS NO NEED FOR H.C. (healthcare) SEE? THERE IS NO NEED FOR MEDICINE'S! (In the Christ mind you do not get sick. If you do not get sick you have no need for medicines or health insurance, or doctors.)

I SEE MORE CRIMES: I SEE A HOUSE ESTATE SCAM'S IN AMERICA. AMERICANS LOST THEIR HOMES DUE TO THE HOUSE ESTATE SCAMS. THE FORECLOSURE OF HOME'S IS DUE TO THE FAC(K)E TRADES OF UNITED AND AMERICAN STOC(K)S IN THE FINANCE SECTOR. I SEE THE CRIME SCENE OF INSIDERS TRADES! I SEE AN INCREASE IN STUDENT FINANCE (student loans), THE HEADS OF AMERICA MADE MORE TENDER (MONEY) FROM THE INTEREST ON STUDENT FINANCE (loans). I SEE THE RISE IN THE USE OF CREDIT CARDS.I SEE INTEREST RATES SOAR ON CREDIT CARDS. I SEE THE TENDER (MONEY) MADE FROM INTEREST RATES.

I SEE "ITT" DOT COM SITES ARE ON THE RISE.

I SEE IN THIS NAME ART THAT "THE FOURTH RIECH" SENDS FIREARMS TO ISREAL AND IRAN. I SEE THE "FOURTH RIECH" FUNDS EACH SIDE'S OF THE ARMIES, SENDS FIREARMS TO EACH SIDE. (Means they fund both sides of the war to keep them going

I CAN SEE AMERICANS ON CAMERA'S. (surveillance cameras everywhere)

A SEER IN CHRIST CAN SEE THE END TIME'S. A SEER CAN SEE A DEFEAT OF THE "FOURTH RIECH" A SEER CAN SEE THE TRIALS OF THE "FOURTH RIECH" FOR THE IN-HUMANE ACTS ON HUMANS OF THE EARTH! A CHRISTED MIND IN

ONE-NESS CAN READ THE MIND OF OTHERS. SHARE THE TRUTH ON THE NET (INTERNET) MEDIA SITES. I SEE THE CRIMES OF THE HEADS OF AMERICA (Presidents). I SEE THE C.I.A. I SEE THE MAFIA I SEE THE HEADS OF E.U. (EUROPE) I SEE THE HEADS OF ISREAL! I CAN SEE THE FREEMASONS OF THE EASTERN STAR; OUR FORE-FATHERS .NO ONE CAN HIDE FROM THE ACTS, THE SCENES, AND THE STORIES IN THIER NAME!

AMEN- ME A NAME AMEN, A CHRISTED MIND ON THE EARTH.

CHRIST COMES TO THE HUMANS ON EARTH; HUMANS D.N.A ON EARTH RE-DOES THE HUMAN MIND. THE CHRIST MIND IS HERE! A TRANSMISSION FROM THE INSIDE-OUT!

NAMASTE!

Channeled June 11 2015 copyright© C H R I S T

JOYCE ANN MORFORD 3 8 18 9 19 20= 77

22 11 44 = 77

THE UNITED STATES OF AMERICA SYNOPSIS

The Name art of The United States of America as you can see contains situations most people are unaware of. As you can see Letters are interchangeable and by using my innate I am able to tap into hidden facts that the name vibration carries.

I can see, hear, and feel the acts, the scene and the story within the name. Everything that you are reading past, present and future are the facts.

Americans have been lied to for a very long time. Some will be in utter shock while reading this. Some may say: It has been right in front of our noses all along.

Every name contains positive and negative vibrations, some more so than others. I am only revealing what I see, hear and feel, to make people aware of the spoken word. Words are things manifested in the flesh, or manifested on the material plane. It doesn't matter if it's a person, place or thing, it has a vibrations and it has an act, scene or story.

I really do not have to say much about this synopsis since all American's have been living under this name vibration for a very long time. I have a gift of revealing, "Yes" of sometimes very negative situations. I do so in order to try to raise the vibration of humanity and its Awareness.

This Awareness is to show the entanglement of everything. We are all entangled together in some form or another. This name art reveals the darkness we have been blinded to; and yet there is still a positive message in this name for all. We unite as one in a crisis. We unite as equal human beings on the Earth to live a life of Joy and Happiness and Peace; with not one of us being above or below the other. All of us have a right to housing, food, medical services etc. There are enough resources for all. You and your birth right to live a descent and FAIR life have been taken away from you from the so called "Elite" of the planet. The greedy one's, the one's that want power over you to enslave you. Just like the Nazi's and the "Third Reich" We have the same continuation now called "The New World Order" or "Fourth Reich" Hitler did not die, he helped create this. It's time for the masses to wake up and fight the darkness.

The darkness of this planet was created by an alien race long ago; we call them the Draco- the draconian. Yes! ET'S exist' you were jut lied to and manipulated for song long. It is time for disclosure.

Many of us channel ET's of a higher agenda. One of Love, co-operation living in unity. The positive ET's want to see and help us evolve. We should embrace them for they are of the creator also. There D.N.A. has surpassed ours for they have been around the block a lot longer than we have. You were only made to be fearful because the Negative ET's feed on your fear. Your Soul is eternal! You have nothing to fear! Love is the source of your power. Maybe it's time for the masses to realize their Ego as the false self it is; for behind the false personality and persona, your true spiritual self is there.

The outcome say's we will win! I truly believe this. Stop them from stealing our money!!!!! Stop them from controlling you for their own material gain, there Super Ego is at work! Like the Name Art say's "It's time to join as one in a crisis and beyond duality consciousness! Namaste

THE NAME ARTS
OR
SCIENCE OF THE ART
MASTER'S EXAMPLES

LEONARDO DI SER PIERO DA VINCI

A DA VINCI CODE

I, A SEER SEE, O DRACONIAN DEVIL, IS LEONARDO DA VINCI!

I A SEER, I CAN SEE LEONARDO IS A SEER.

LEONARDO IS A DIVINE ONE, I CAN SEE LEONARDO CAN AND DOES DIVINE IN CODE, DIVINE IN COLORS. I A SEER, I CAN DECODE LEONARDO.SEE D.N.A AND R.N.A. ARE CODES!

I SEE NO LIES IN LEONARDO! I CAN READ LEONARDO'S CODES! A SEER EAR'S RECIEVES SONAR, RADAR! A SEER CAN SEE D.N.A AND R.N.A AS A RADIO RECIEVER! A SEER PIERCES A VEIL! (Subconscious mind is the pierced veil) D.N.A AND R.N.A IS A DIVINE SEED! LEONARDO DA VINCI'S D.N.A. DANCES, D.N.A. IS ALIVE!

I A SEER; I SEE A SEER IN LEONARDO DA VINCI.

LEONARDO DA VINCI IS A SEER.

LEONARDO CAN SEE PI AS A DIVINE SPIRAL IN D.N.A.!

LEONARDO RECEIVES A DIVINE VISION- A PINE CONE- PINEAL. PERCIEVES A PINEAL ENDOCRINE AS A RADIO, AS A DIVINE RADIO RECEIVER! LEONARDO PERCEIVES NIRVANA. LEONARDO PERCIEVES AN EAR, LEONARDO"S EARS RECIEVES SONAR! PERCIEVES DNA AND R.N.A AS A RADIO AND RECEIVER! D.N.A AND R.NA IS A DIVINE RADIO RECEIVER! LEONARDO CAN SEE SACRED COLORS IN AN INNER CAVE! (MEANING ,HE SEE CHAKRA COLORS WITHIN HIMSELF)

LEONARDO CAN SEE AS ONE. LEONARDO CAN SEE A DOOR OPENS AND NO VEILS ARE SEEN. (Conscious mind and sub conscious mind are one, see no separation) LEONARDO CAN SEE A CLEAR VISION! LEONARDO PIERCES A VEIL;

A DIVINE ONE NEAR LEONARDO RAISES A VEIL. LEONARDO CAN READ A CODE. LEONARDO ENCODES A CODE IN COLOR ON CANVAS. LEONARDO COLORS AN OVAL VESICA PISCIS OVER A DIVINE LORD; OVER DIVINE DISCIPLES. SEE A VESICA PISCIS IS A SACRED CODE.

A VISICA PISCIS IS SACRED.

LEONARDO IS A SEVEN. LEONARDO CAN SEE SEVEN AND ELEVEN AS DIVINE. SEVEN AND ELEVEN AS PI SPIRAL! LEONARDO IS EVEN! (balanced male/female energies) LEONARDO CAN SEE EVE IN ELEVEN AND SEVEN.LEONARDO SEE EVEN IN SEVEN AND ELEVEN.

LEONARDO CAN SEE AS ONE. LEONARDO CAN SEE A LIE. AN EAR NEAR A RADIO.LEONARDO EARS NEAR A RADIO= (PSYCHIC) =PINEAL

LEONARDO IS NEAR A DIVINE ONE IN D.N.A. LEONARDO'S EARS NEAR A DIVINE VOICE IN D.N.A. AN INNER VOICE! A VOICE IS IN SOPRANO. A SOPRANO VOICE IS EVE! LEONARDO SEE EVE AS SACRED! LEONARDO CAN SEE NO EVIL IN EVE!

LEONARDO IS IN A SACRED ORDER IN PARIS.

SEE PRIORI DE SION.PRIORI DE SION CODES ARE A PELICAN, A DOVE, A VINE, A PEAR, A RIVER, A ROSE, AND A RAVEN.LEONARDO ARRIVES= (BORN) IN APRIL.

LEONARDO IS AN ARIES. LEONARDO CAN SEE A ROSE AS SACRED. LEONARDO CAN SEE A ROSE LINE. LEONARDO CAN SEE AN ANCIENT ROSE LINE IN PARIS.

LEONARDO PERCEIVES VENERIS= (FEMALE PUBIS) PENIS, OVARIES AS SACRED, AS DIVINE.

LEONARDO RAPES A CADAVER, (rape meaning opens, cuts through) CAN SEE VEINS AS A DIVINE RIVER, A RED RIVER A DIVINE VOICE. (a red river is sacred blood that D, N.A and R.N.A flows in GENES) LEONARDO; AN EAR NEAR A DIVINE RED RIVER, A DIVINE VOICE! LEONARDO CAN SEE D.N.A IN A DIVINE RED RIVER! (the blood)

LEONARDO DA VINCI CAN DO CODES IN REVERSE.CAN READ IN REVERSE, LEONARDO CAN SEE IN REVERSE.

LEONARDO COLORS IN PENCIL, (sketches) I SEE AIRPLANES IN PENCIL.

LEONARDO; DIVINES IN OIL ON CANVAS, ON AN EASEL. LEONARDO REDOES A DISCIPLE ON CANVAS IN OIL! SEE A DISCIPLE NEAR A SAVIOR, A LORD. SEE A "V"; SEE A SAVIOR AND AN EVE DINE AS ONE! PEER CLOSER AND SEE AN EVE AS A DISCIPLE! LEONARDO REVEALS A CARDINAL, A DIOCESE LIES ON CANVAS! A DIOCESE ORDERS SILENCE! LEONARDO'S LIPS ARE SEALED! LEONARDO IS A CLEVER ONE, LEAVES CODES ON CANVAS!

A SEER SEE AN OPEN DOOR, SEE AN ENDOSCOPE (CAMERA), AN ENDOSCOPE CAN SEE A CANVAS IS REAL! I SEE A VASARI.

I; A SEER CAN SEE LEONARDO'S EAR IS NEAR AN ALIEN RACE IN D.N.A. AN ALIEN RACE IS A PLEIADIAN RACE. AN ALIEN RACE IS NINE-ENNEAD. LEONARDO'S EAR'S NEAR NINE-ENNEAD!

LEONARDO'S D.N.A IS ALIEN! LEONARDO'S D.N.A IS PLEIDIAN. PLEIDIANS INSPIRE LEONARDO DA VINCI. NINE- ENNEAD INSPIRE LEONARDO!

LEONARDO LEAVES A SECRET CODE IN COLOR ON CANVAS! A CANON (of) ONE RACE!

LEONARDO DI SER PIERO DA VINCI SYNOPSIS

The first name art that I would like to present here is an art master Leonardo Da Vinci although using his full name. Here I will explain the art in this synopsis. When I am doing the name art it is as if I was accessing the consciousness of the one carrying the name. I am the seer viewing the consciousness through his name. This name art tells of Leonardo Davinci is a seer which is (clairvoyant, clairaudient he is psychic. He uses code to reveal his revelations because of the church. In those days his findings would have been blasphemy. So his paintings reveal a lot of esoteric doctrine in the colors and the paintings of religious nature. A seer ears receive sonar, radar which mean he is clairaudient and can hear messages within him. It reveals the D.N.A. within him. His DNA is activated. His D.N.A.is a radio and receiver. This means D.N.A. emits information and it also can receive information.

His name art explains of his discovery of using PI in his art work and PI being a divine ratio; just as the numbers 7 and 11 are divine numbers. Leonardo's name reveals the vesica piscis that he used in his drawings and art work. He uses the vesica piscis as a halo over the divine in his paintings. It also tells of the sacredness of the 2 circles overlapping each other as a womb for birth. Male and Female coming to create not only the birth of the human form, but also the birth of creation, even an idea which all springs forth.

The veil that Leonardo pierces is his subconscious mind; and the divine seed ignites in his D.N.A. He is seeing with his heavenly site. His name reveals he is seeing with his pineal gland that has been activated.

It reveals that our D.N.A. is a radio and receiver. Our D.N.A. can send and receive messages; which are also godly in nature; the messages from our divine self; or our higher self.

His name also reveals his presence in a secret or in Paris the Priori De Sion and the symbols they used esoterically.

His name reveals when he was born into the Earth. It reveals an Aries born in April.

Leonardo reveals that blood is divine and Leonardo worked on cadavers. It reveals how Leonardo could read in reverse and used reverse writing. Leonardo reveals how the church lies and he reveals this in code with his paintings and writing.

Leonardo's ear near an alien race depicts that his information was from a higher nature an alien race giving him knowledge.

MICHAEL ANGELO DI LODIVICO BUONARRATI

OM!

ME A NAME AMEN! A TRUE ART!

I AM A SEER ON EARTH!

I AM AN ANGEL, AN ANGEL ON EARTH!

AN ANGEL AND I ARE ONE!

I CAN NARRATE AN ANCIENT ART!

I NARRATE AN ANGEL ART!

I BRING A LIGHT TO THE EARTH IN ART!

I AM THE REAL ME- NO EGO IN ME ON EARTH.

I READ MICHAEL ANGELO'S D.N.A. a NAME THE SAME AS D.N.A.!

I AM MICHEALANGELO BUONARROTI. A BRILLIANT MAN ON EARTH! A BRILLIANT ANGEL ON EARTH! I LEAVE A RARE ART ON EARTH.

I AM BORN TO A MOTHER NAME NERI. I AM BORN IN CAPRESE. I AM A SEER HERE ON EARTH.

(He had intuitive capabilities) I CAN HEAR AN ANGELS VOICE INSIDE ME.ICAN HERE THEM FROM THE OTHER SIDE OF THE VEIL

I AM NEAR AN ARCH ANGEL. AN ARCH ANGEL NAME GABRIEL.THE ARCH ANGEL CAME TO ME IN A DREAM. I AM NEAR AN EAR, NEAR A NOUMENON= (SOUL, MIND, AN INTELLECTUAL THING AS IT IS IN ITSELF, NOT KNOWN AS IT IS THROUGH PERCEPTION), I AM NEAR A MUSE. I AM NEAR THE DEAMON (HOLY GUARDIAN ANGEL) OF ME! MY EARS RING, MY EARS SING. MICHEAL ANGELO CAN SEE THERE IS ONE REAL THING ON EARTH LOVE! AN ANGEL CAN TELL NO LIES, AN ANGELS HEART CANNOT HATE. THE HUMAN MIND OF MEN, THE EGO DOES THESE THINGS!

I AM A MAN ON EARTH, I AM AN ANGEL ON EARTH, AND I AM A LIGHT BEING ON EARTH! I AM TORN ON EARTH, I AM TORN IN RELIGION, I AM AN ILLUSION ON EARTH. ME A NAME A DREAM A VIBRATION (OF) LIGHT! THE LIGHT IN ME HAS NO NAME! ON EARTH I AM THE VIBRATION (OF) THE NAME, HUMAN, NOT THE REAL ME! I HAVE A REAL NAME AN ANGELIC NAME A STONG VIBRATION (OF) LIGHT! I CAN SEE AND HEAR INVISIBLE THINGS! SHUT OUT THE MIND, LET THE SUN, THE REAL ME SHINE! I AM AN ANGEL ON EARTH. HUMANS DO NOT SEE THE GODS IN THEM. I LAUGH AT THEM, SEE IT ON OIL, SEE IT ON CANVAS, SEE IT IN STONE, AND SEE IT ON THE CEILING! HA-HA! SEE THE ANGELS

ON THE CEILING IN THE CENTRE (city) VATICAN IN ROME. I CREATE IN OIL ON THE CEILING. I DIVIDE THE CEILING IN TO CREATE ANGELIC ART. I SEE IN OIL (PAINT); "THE CREATION'S (of) HEAVEN AND EARTH". I SEE; "THE CREATION (of) ADAM." I SEE; "CHRIST COMES AND THE SALVATION (of) ISREAL." I SEE "NUDE YOUTHS.I SEE- NOAH AND SEER'S. I A SEER CAN SEE LEONARDO DA VINCI IS THERE AT THE SAME TIME; SAME ERA!

I GO TO LEARN LATIN. I GO TO A MAN NAME GALEOTA IN URBINO TO LEARN LATIN.

I MICHEALANGELO RATHER LEARN ART. I TURN A CAREER TO LEARN ART! I CAN IMAGE IT ALL IN THE ONE HEART! ALL ARE ONE! I IMAGE A BALANCE A GRACE IN THE HUMAN ART, IN THE HUMAN HEART!

I IMAGE A HUMAN ARM, A HUMAN LEG. I: IMAGE A HUMAN HEART. I IMAGE THE HUMAN BONE. I GO TO A MORGUE TO LEARN.

I CUT ON A HUMAN TO LEARN AN ART! I LEARN ALL ARE ONE ON EARTH! I LEARN MEN ARE BORN BARE, NOT IN CLOTH.

I MICHEALANGELO CAN TURN MARBLE INTO ART! I NOTCH OUT, I CUT, AN ETCH A HUMAN MALE IN MARBLE. I CUT AND ETCH AN ANGEL ART. I CREATE A HUMAN MALE IN MARBLE. I BRING ABOUT AN ANGEL ART TO THE EARTH.

I LEARN A BALANCE IN NATURE. I CAN CREATE NATURE IN MARBLE.

A MEAN MAN RAGE AND RULE. A MEAN MAN CAME TO BURN ALL THE ART NOT OF THERE RELIGION.

I GO TO ROME TO BE ABLE TO UN EARTH AN ANCIENT RUIN. I CUT OUT AN ETCH A LARGER IMAGE IN MARBLE IN BARGELLO. I NAME IT (BACCHUS).I MICHEALANGELO CAN RAGE. I HARBOR AN ANGER IN ME! I AM A RUDE MAN. I HARBOR NO MANNER IN ME. NO ONE CAN TOUCH ME! NO ONE IS NEAR ME!

A THEOLOGICAL REILGIOUS MAN RENDERS MICHEALANGELO A CARTE BLANCHE= (UNRESTRICTED POWER TO ACT AT ONES OWN DESCRETION). MICHEALANGELO COLOR A MURAL ON A CEILING IN ROME.IN ROME, MICHEALANGELO CREATE ART IN A TOMB. A HUMAN IN MARBLE, A GREAT MAN IN AN ARMCHAIR ERECT IN A NICHE.

MICHEALANGELO; HE MEET ANOTHER GREAT MAN IN ART! A MAN NAMED LEONARDO DA VINCI.

A MAGLIABECHIAN AUTHOR CAN TELL AN AMUSING TALE! MICHEALANGELO RAN INTO LEONARDO; IN ARROGANCE HE TELLS LEON ARDO;HE RAN IN MILAN. A RELIGIOUS MAN HIRE BOTH OF THEM TO COLOR A MIRROR MURAL.NO MORE THAN A CARTOON IS CREATED.

MICHEALANGELO MET A GIRL NAME CALONNA.

AN ANGEL GIRL CAN HEAL A HEART. CAN HEAL MICHEALANGELO HEART!

ME, I AM A NAME, AN ART. THE ONE BREATH, THE ONE ECHO ON EARTH. I AM ANIMA, I AM ANIMUS, I' AM ONE. I AM A TEACHER IN THE MATERIAL EARTH. I BRING A LIGHT TO THE IGNORANT ON THE EARTH. I AM A LIGHT BEING, AN ORACLE ON EARTH!

ME A NAME AMEN!

AUTHOR: Joyce Morford, and her Monad June 2003 Copyright ©

MICHAEL ANGELO DI LODIVICO BUONARRATI SYNOPSIS

As a Seer, I have my ego aside in order to channel the name. The ancient art I talk about is called Nomancy. Nomancy is divining a person's life by the letters that form their name. When your ego is put aside your angel self can come forward and channel the art.

Michael Angelo was in touch with his Angel self, his creative self; his muse. When he was sculpting his non-ego side took over to create the beautiful sculptures you see today. He was in touch with an angel; an Arch Angel Gabriel. Michael Angelo was a brilliant man because he was in touch with his genius. This tells me he had conversations with angelic beings from the inside of himself. He was in touch with his holy guardian angel; his holy self. Michael Angelo was able to communicate with the other side of the veil! Michael Angelo was a self-aware human being in his time. He realized that the Earth and his life was an illusion. HE realized he was a light being in a human form. In those days Religion and the church ruled and any of this would have been blasphemy! So he lived his life torn knowing the real truth. As you see his name says so. He was also close to his mother. His mother's name is in his name.

His name also conveys when he was born and where.

He talks about learning Latin and the name of the person that taught him.

His name reveals that he knew Leonardo da Vinci and that he was not fond of him.

His name reveals a girl that he was interested in. His name reveals the purpose why he came. This is no accident. He was a greatest sculpture of his time.

THE NAME ARTS OR SCIENCE OF MISSING PEOPLE

NATALEE ANN HOLLOWAY

WE TELL A TAN ONE; AN ANN, TO TELL ALL A NEW LAW! AN ANN TELL ALL THE NEW LAW!

WE TELL ANN A LOT! AN ANN, AN OWL.

NATALEE WON AN "A"; IN ALL.NATALEE WANT A YELLOW HAT, A YELLOW HAT ON LAY-AWAY.

NATALEE- A NEW ONE IN TOWN.

NATALEE A HOLY ONE, NATALEE HATE NO ONE.NATALEE WANT TO HEAL

NATALEE NOT TALL AT ALL. (SHORT)

WHO? WHEN? WHY? AN HOW? NATALEE WENT AWAY. NATALEE ACRONYM=WAT= (WITH OUT A TRACE) NATALEE WENT AWOL= (ABSENT WITHOUT LEAVE). NATALEE WENT TO A NEW TOWN ON AAL= (AMERICAN AIR LINES)

NATALEE ON A TELLY. = (TELEVISION)

A TOWN WENT ALL YELLOW. THEY TYE YELLOW ON AN ANTENNA (RIBBONS)

NATALEE A WEALTHY ONE.

NATALEE; YELL, Y'ALL A LOT! NATALEE YELL HELLO!

A TALL ONE, A LEAN ONE, TELL A WHALE TALE. A TALE NOT TALLY=

(DOESN'T ADD UP)

NATO ON THE WAY!

THE LAW ONTO NATALEE. THE LAW ON THE WAY!

NATALEE LOAN A LOLLY, LOOT (MONEY)

NATALEE NOT AT ATA= (AIRPORT TRAFFIC AREA)

NATALEE NOT ALL ALONE ON THE TOWN. NATALEE NOT ALONE AT ALL.

NATALEE NOT ALONE AT A HOTEL. WE WANT NATALEE ALONE. NATALEE NOT AT HOTEL LATE. NATALEE ATE ATOLE AT A HOTEL. NATALEE NOT AT A HOLLOW HOTEL. NATALEE A TOY TO A TALL ONE. WE TELL NATALEE NOT TO YELL.TWO TAN ONE; A HE, TELL NATALEE NOT TELL. NATALEE WENT AWAY.. NATALEE YELL THE WHOLE WAY. NATALEE WENT ANY WAY. NATALEE LEAN ON A HALLWAY WALL, AN ALLEY WALL, A YELLOW WALL. WE WANT NATALEE, NATALEE A HOT ONE. NATALEE A TEEN.

NO WALLET ON NATALEE. NATALEE WANT A TAN. THEY WHEEL NATALEE AWAY. A TALL ONE EYE NATALEE. A TALL ONE AN TWO TAN ONE. TWO TAN ONE TELL NO ONE! A TALL ONE TELL

NO ONE, HE WANT NATALEE. NATALEE NOT WET AT ALL.NATALIE LAY ON A TOWEL. A TALL ONE NOT TAN, LAY NATALEE ON A TOWEL. A TALL ONE A HEEL! ONE TAN, ONE NOT TAN LAY NATALEE ON A TOWEL. WE LAY NATALEE ON A TOWEL. A TALL ONE NOT TO TELL. NATALEE YELL NO!!

NATALEE- NOT WELL AT ALL. A LETHAL ALE (DRINK) A TEA! NATALEE WENT LOONEY (?) NATALEE LAY LOW. A TALL ONE LAY LOW.A TALL ONE LAY NATALEE LOW ON A HEEL= (AFTER END OF A SHIPS KEEL, LOWER END OF A MAST.) A TALL ONE TYE NATALEE TO A HEEL. NATALEE ON A YAWL= (A SAILING VESSELL WITH 2 MASTS, A SMALL BOAT CRUED BY OARS.)THEY TOW NATALEE ON A YAWL.THEY ENWALL NATALLE.NATALEE LYE ON A WELL WALL. WELL= (AN ENCLOSE COMPARTMENT OR RECESSED AREA USED FOR STORAGE ON A BOAT OR VESSELL) THEY LAY NATALEE ENTO A HOLLOW HOLE! A HOLLOW HOLE A HOT HOLE. NATALEE WENT ALL THE WAY TO HELL A HOT HOLE. A HOLLOW HOLE ON A YAWL. THEY WONT LET NATALEE TELE= (TELEPHONE) NO ONE! A TALL ONE LEAN NATALEE ON A WALL. NATALEE ON A YELLOW WALL. HOLLOW HOLE! A HOLLOW HOLE HOT, A HOLLOW HOLE HELL.A HEAT A HOT ONE!

THEY TOW NATALEE TO ATOLL= (AN ISLAND CONSISTING OF A CIRCULAR CORAL REEF.) THEY

TOW NATALEE ALL THE WAY TO WAAL= (LOWER RHINE IN SOUTHERN NETHERLANDS)

NATALEE WENT TO ALTO, NATALEE WENT TO AYO.NATALEE ON THE WAY TO A YELLOW HALLOW ALTO (VISTA CHURCH) NATALEE ON THE WAY TO AYO (ROCK FORMATIONS)

HAWALA= (AN UNDERGROUND BANKING SYSTEM BASED ON TRUST WHERE MONEY CAN BE MADE AVAILABLE INTERNATIONALLY WITHOUT THERE BEING A RECORD OF THE TRANSACTION.)(TERRORIST USES THIS SYSTEM)

A TALL ONE TOW NATALEE AWAY ON WHEEL(S)

THE LAW TOW AWAY WHEEL(S)

NATALEE ON A LOT; A NET ON NATALEE.

LOY= (A LONG NARROW SPADE FOR STONEY LANDS)

NATALEE NOT WHOLE AT ALL, NATALEE EN TWO

THEY LOWER NATALEE EN-TO A HOT HOLE,

A HOT HOLE A HELL! (BURNED HER BODY). NATALEE NOT HOT NOW: NATALEE AWAY. (PAST AWAY). NATALEE WENT AWAY NOT ON (HER) OWN! NATALEE WENT AWAY ON THE TENTH TEE.

Author : J.A.M. copyrights March 31, 2005 ©:

JOREN ANDREAS PETRUS VAN DER SLOOT

READ A RARE ART! A SEER'S EARS NEAR SONAR, NEAR RADAR, A SEER READS D.N.A. A SEER'S EARS NEAR NATALEE'S SOUL, NEAR JORENS SOUL. A SEER ANN REVEALS A RARE ART! A RARE ART REVEAL A LOT! A SEER REVEAL'S LETTERS LET'S A SEER SEE! LETTERS ARE SAME AS D.N.A!

A SEER READ'S A RED ALERT! A RED ALERT ON ORANJESTAD!

A SEER READ'S JOREN ANDREAS PETRUS VAN DER SLOOT- READ'S JOREN'S D.N.A.

A SEER READ'S AND SEE'S JOREN ON ORANJESTAD!

JOREN TOLD A TALE! JOREN'S TALE'S DO NOT ADD UP! DO NOT TRUST JOREN OR JOREN'S DAD! DO NOT TRUST JOREN'S PARTNER'S. JOREN'S PARTNERS TALES NOT TRUE! JOREN'S TALES NOT TRUE!

A SEER SEE'S JORENS NEED'S A PLOT; JOREN NEEDS A PLAN. SEE JOREN'S PLOTS AND PLANS!

A SEER SEE'S JOREN AS A SOT= (DRUNKARD)

A SEER SEE'S; NATALEE'S PARENTS ON ORANJESTAD.

SEE NATALEE'S PARENTS AT NOORD.

A SEER SEE'S NATALEE'S PARENTS PLEAD A DESPERATE PLEA!

SEE NATALEE'S POSTER'S ON ORENJESTAD, SEE'S US DOLLARS ON NATALLEE'S POSTER'S.

A SEER SEE'S NO PURSE ON NATALEE; NATALEE NOT ON A PLANE TO US.

A SEER SEE'S JOREN AND NATALEE ARE ON A DATE. JOREN RANT AND RAVE OVER NATALEE.A DATE NOT A REAL ONE. JOREN DOES NOT LOVE NATALEE; JOREN JUST USES NATALEE.

JOREN AND NATALEE ARE SEEN AT A TAVERN A SALOON. (BAR) NEAR A SLOT= (MACHINE), NEAR LOTTO.JOREN ORDERS NATALEE A TEA (A DRINK); A SEER SEE DOPE ENTERS A TEA (DRINK) SEE AN OVERDOSE, SEE NATALEE AS DOPED UP. JOREN RODE ON A REAR SEAT (BACK SEAT) NEAR NATALEE. DROVE A VAN OR A LAND ROVER ON ORANJESTAD. JOREN DROVE A RED JEEP. JORENS PARTNERS ARE TAN (DARK SKIN). JORENS PARTNERS DROVE NATLALEE AND JOREN TO A TAVERN A SALOON (BAR). JORENS PARTNERS DROVE NATALEE AND JOREN TO A SAND AREA.(A BEACH) SEE'S NATALEE RAN, NATALEE RUN'S . SEE A JEEP RUNS OVER

NATALEE. (RAN HER DOWN)? SEE JOREN AND NATALEE NEAR A RIVER.

JOREN SLANDERS NATALEE (FLASE STATEMENTS ABOUT HER)

JOREN STEAL'S NATALEE ON A RESORT ON ORENJESTAD.(KIDNAPPS)

JOREN'S A ROOSTER= (A MALE CONSIDERED AS COCKY) SEE JOREN PLANT'S HIS LIPS ON NATALEE (A KISS) A SEER SEE DATE RAPE.SEE A TORN DRESS JOREN RAPES NATALEE ON SAND. SEE SAND AND SEA (A BEACH)

JOREN EVADE'S NATALEE NEAR A SEA. (SEASHORE).

JOREN'S SEEN AT A SEVEN ELEVEN LATE OR JORENS OUT LATE PAST ELEVEN.JOREN SEE'S A SUNRISE PAST SEVEN.

SEE A PLOT, JORENS'A SON TO AN EASTERN SLAVE TRADER, JOREN'S A SON TO A SLAVE VENDOR, JOREN DROVE NATALEE TO AN EASTERN SLAVE TRADER. JOREN TRADED NATALEE TO A SLAVE VENDOR, AN EASTERN SLAVE TRADER.

JOREN REVEALS AN OLDER ONE SOON! SEE AN OLDER VAN DER SLOOT, A SR. VAN DER SLOOT! SEE PAUL VAN DER SLOOT! PAUL VAN DER SLOOT A VENAL ONE!= (CAPABLE OF BEING CORRUPTED, CORRUPT JUDGES,POLICE OFFICERS FOR MONEY OR A PRICE). A SEER SEE'S JORENS

ARRESTED AND AN OLDER VAN DER SLOOT PAUL IS ARRESTED ON ORANJESTAD! A SEER SEE JORENS PARTNER'S ARE ARRESTED ON ORANJESTAD, JORENS PARTNERS ARE NOT RULED OUT. SEE PARTNERS ARE RELEASED.

JOREN REVEAL'S NATALEE NOT ON LAND, NATALEE AT SEA. SEE NATALEE ON A SEA VESSEL. A SEER SEE'S NATALEE'S LOST AT SEA. JOREN SAT NATALLE ON A STERN,; ON A SEA VESSEL. A SEER SEE'S JOREN LEAVES NATALEE ON A SEA VESSEL AT SEA.

SEE TOOL'S USED, A SPADE, ROPE, PLASTER, AND TARP. SEE JOREN LOOSENS A NOOSE; SEE A ROPE ON A SPOOL OR REEL.

AN OLDER VAN DER SLOOT PAUL A SOLON= (A RESPECTED LEADER, LEADER IN NATIONAL AND INTERNATIONAL AFFAIRS).

JOREN DOES NOT SAVE NATALEE. A SEER SEE JOREN OARS A TENDER TO A SEA VESSEL. A SEER SEE'S OARS AND PADDLES.

JOREN REVEALS A LOT SOONER OR LATER. SEE'S JOREN IS UNDER PRESSURE, UNDER DURESS. A SEER SEE'S NATALEE'S PARENTS PLEADED NOT TO LET JOREN'S PARTNERS LEAVE ORANJESTAD. NATALEE'S PARENT 'S PLEADED TO U.S. SENATORS. A SEER SEE'S PLANES ARE SENT TO ORANJESTAD. PLANES USE SONAR AND RADAR.SEE LENSES AND PODS UNDER A PLANE. PLANES DO NOT SEE NATALEE.A SEER SEE'S

NATALEE, A SEER'S EAR'S NEAR NATALEE'S SOUL!

NATALEE RAN AND LOST A SANDAL. A SEER SEE'S NO PULSE ON NATALEE! A SEER SEE'S NATALEE AS DEAD! SEE'S NATALEE SLOOPED OVER, NO POSTURE.

A SEER SEE'S AN ARSON ON ORANJESTAD. SEE AN OUT DOOR OVEN (CREMATORY)A STEEL VAULT. SEE NATALEE AS SEARED AND ROASTED. (CREMATED)? SEE NATALEE'S PART'S ARE SOOT! SEE NATALEE'S PARTS AS SOOT ENTER'S A VASE OR JAR. DO D.N.A ON SOOT, ON DUST. A SEER SEE'S AN UNJUST PLAN ON ORANJESTAD! A SEER SEE'S JOREN DOES IT IN PERU! A SEER SEE'S JOREN IN PERU.SEE'S JOREN'S ARRESTED IN PERU! (JOREN IS IN PERU AND HE MURDERS ANOTHER WOMAN!)

SEE JOREN DOES NOT LOVE; JOREN NEEDS TO LEARN TO LOVE. (JOREN IS A BIT OF A SOCIALPATH, HAS NO EMPATHEY FOR OTHERS)A SEER READ'S JOREN LETTERS- LETTERS = LETS A SEER SEE. LETTERS ARE THE (SAME) AS D.N.A!

AUTHOR J.A.M.

SYNOPSIS: NATALEE ANN HOLLOWAY AND JOREN ANDREAS PETRUS VAN DER SLOOT

When I was channeling these 2 name arts I was amazed of the entanglement of the 2 parties involved.

The first name Natalee Ann Holloway foretells of her being a straight A student. Her name tells of the events and the fact that she would disappear without a trace.

Her name tells of a Tall one and two tan ones which are Joren and his 2 friends.

In Joren's name Natalee's name is entangled with his. Joren's name gives a lot more detail of what happened. In Joren's name it states Oranjestad the place where it occurred so it is no accident that it happened. His name reveals do not trust him. His name reveals all the plots and plans that occurred in the acts the scenes the story. Joren's father

Paulus is implicated in his name. His 2 friends are revealed in his name. I channeled that Natalee's remains would never be found.

At the end of the channel I got "See Joren in Peru."

It is too bad that law enforcement agencies do not look past their visual eyes. Sometimes they need to trust a Seer! Law enforcement should start working with Seer's who have access to information that the normal 5 senses of the human being cannot see, cannot hear. I believe there would be less mistakes of the innocent and guilty.

By only using the letter of the name the story unfolds.

I see a macro code in everyone's name. Their names foretells of the situations they would encounter in their life time. Usually the entangled party is in the others name, including friends you have, family members ect. Sometimes it is not a happy ending; although I do believe in set-ups before we came to Earth.

I believe in this formula for divination. In my D.N.A. this gift has always been with me from previous lifetimes. This truth is revealed in my own name art. Sometimes the truth is accessed by intuition alone. One cannot always be certain of just the facts in front of you to determine if someone is guilty or not. I can attest to this because in my family there was a murder disguised as an illness. This person thinks they got away with it. The other side of the veil is accessible if you only listen.

Here is an old saying: The sun rises in the east and sets in the West, the other side knows best!

JESSICA MARIE LUNSFORD

OM!

ME A NAME AMEN!

I A SEER CAN SEE AN OMEN IN A NAME! I AM A SEER; I READ A NAME!

I READ JESSICA MARIE LUNSFORD NAME!

I SEE A NAME SAME AS D.N.A. AND R.N.A.I SEE D.N.A IS A RADIO ON!

I A SEER CAN SEE JESSICA IS NEAR OUR LORD JESUS, NO FEAR!

JESSICA SEE'S JESUS IN A DREAM!

I SEE NO SIN IN JESSICA.

I SEE JESSICAS IS ON CAMERA ON FILM, ON A RADIO.

A SEER CAN SEE MAJOR CODES AND CLUES IN JESSICA'S NAME! A SEER CAN SEE A CRIME SCENE IN JESSICA'S NAME. A SEER CAN DEFACE A NAME. I SEE A CRIME SCENE UNFOLDS IN JESSICA'S NAME.

I A SEER CAN SEE JESSICA'S MISSION IN LIFE; JESSICAS ESSENCE, JESSICA'S LIFE FORCE! I SEE A CAMEO OF JESSICAS LIFE!

I CAN SEE MR. LUNSFORD AND MR LUNSFORD'S MA RAISES JESSICA.

I CAN SEE A RAILROAD NEAR A FARM.

I CAN JESSICA ON A FOUR ACRE FARM. I CAN SEE JESSICA RIDE'S ON A MARE, (A HORSE). I CAN SEE JESSICA IS NEAR FARM ANIMALS. I SEE JESSICA IS FOND OF ANIMALS.

I CAN SEE JESSICA CARRIES A MARINE ANIMAL; A SEA ANIMAL. IS A MARINE ANIMAL REAL? NO! (STUFFED)

A MARINE ANIMAL USES SONAR. I CAN SEE A SEER'S EARS USE SONAR SOUND. (TO HEAR)

I CAN SEE ON JESSICA'S DRESSER, A FRAME (PICTURE) OF JESSICA AN MR LUNSFORD.

I CAN SEE JESSICA RIDES IN A RED CAR.I CAN SEE MR LUNSFORD RACES CARS OR A FRIEND OF MR LUNSFORD RACES CARS. I SEE A NASCAR RACE. I CAN SEE A MAN RACES CARS IN A CAR CLASS OF ONE OR FOUR.

A MEAN MAN COMES FOR JESSICA IN FLORIDA. A MEAN MAN FLEES A CRIME SCENE. I SEE JESSICA IS IN DREAM LAND.A MEAN MAN'S NAME IS JON (JOHN).

CAN JESSICA SCREAM? NO, CAN JESSICA SEE? NO! JESSICA IS UNDER SAND (DIRT BURY UNDER GROUND)

I SEE JESSICA'S NAME FOR CLUES FOR CODES, FOR SOURCE INFO!I CAN SEE LEADS IN JESSICA'S NAME!

I SEE FEDERAL MEN ISSUE A RED CODE (ALERT) FOR JESSICA. I READ: A MAN NAMED JOE SEE AN EMAIL, AN EMAIL COMES FOR JOE. A MAN NAMED JOE IS NEAR FEDERAL MEN. A MAN NAMED JOE CAN SEE A SEER CAN READ A NAME!

JESSICA'S CASE IS CLASSFIED UNDER ORDERS OF FEDERAL MEN;UNDER ORDERS OF C.I.A. MEN. NO ONE CAN SEE JESSICA'S NAME.

A MEAN MAN COMES FOR JESSICA IN FLORIDA. I CAN SEE JESSICA IS SCARED! JESSICA IS AFRAID! A MAN NAMED JON CAME IN AND FORCED JESSICA INSIDE A CAR.

JESSICA CAN SEE A MAN FRIEND IS A MEAN MAN.

(A MAN NEAR JESSICA PRETENDS HE IS A FRIEND, BE -FRIENDS HER)

I SEE A LOT OF CARS ON LAND NEAR JESSICA.(HOME?)

I SEE A USED CAR SCAM, OR USED CAR SALES, IS A USED CAR SALESMAN. OR A MAN LEASES USED CARS. A CAR SALESMAN CAR'S ARE LEMONS. I SEE A FORD, SEE A CAMARO, SEE A NISSAN, SEE A SUNFIRE, SEE A CAR NAMED ARIES (DODGE), I

SEE A FORD RAM, SEE A SONOMA, A CORSICA, I SEE LINES OF CARS, I SEE NO RIMS ON A CAR.I ALSO CAN SEE SOMEONE CLEANS CARS (DETALIER), I SEE SUDS (SOAP). I SEE SOMEONE CLEANS CARS FOR FREEDOM FORD? ERNIES FORD DEALER.

I SEE SOEMONES SON? I SEE A SON OF AN OLDER FORMER FRIEND OF MR LUNSFORD. I SEE A MEAN MAN CARRIES JESSICS IN A CAR.

I CAN SEE JESSICA SCREAMS FOR MR LUNSFORD, SCREAMS FOR MR LUSNFORD MA! JESSICA CRIES FOR A MA!

I CAN SEE DUAL SCAMS. I CAN SEE CRIMINAL MEN IN JESSICA'A FAMILIES RUN A SCAM UNDER OUR NOSES! NO ONE IN JESSICA'S FAMILIES READ'S JESSICA'S FORUMS, I CAN SEE RUMORS SURFACE IN JESSICA'S CASE. SOMEONE IS FRAMED ; A MAN IS FRAMED. I SEE IN JESSICA'S NAME; JESSICA'S FAMILIES ARE CRIMINAL'S. I SEE CRIMINAL RECORDS IN JESSICA'S NAME. I SEE MR LUNSFORD CRIMINAL RECORDS. I CAN SEE MR LUNSFORD CRIMINAL RECORDS IN CAROLINA.I CAN SEE A FAMILIES JAIL RECORDS.

I CAN SEE LARCENIES=(THE UNLAWFUL TAKING AND CARRYING AWAY OF PERSONAL PROPERTY WITH THE INTENT TO DEPRIVE THE RIGHTFUL OWNER OF IT PERMANTLEY)

I CAN SEE AN OLDER MR LUNSFORD (FATHER) CRIMINAL RECORDS IN INDIANA, SEE ALSO FLORIDA. SEE JESSICA'S SECOND COUSIN. ARE

MEN IN JESSICA'S FAMILIES REFORMED CRIMINAL'S? NO!

I CAN SEE A CROSS ROADS IN JESSICA'A CASE!

I CAN SEE FEDERAL MEN AMASS A SECOND SCAN OF AN AREA; SCOURS AN AREA.

I SEE MAIL-MEN SEND FLIER'S AROUND FOR MILES, JESSICA'S INFO IS ON FLIER'S!

I CAN SEE A MEAN MAN NAMED JON IS FAMILIAR IN AREA. (PEOPLE KNOW HIM)

I SEE A MAN NAMED JON (JOHN) ADORN'S A FACIAL SCAR AND A ROMAN NOSE. I SEE A FACIAL MOLE. I SEE A MAN ADORNS A LEO ON A FORE- ARM (TATOO).

I CAN SEE A MAN NAMED JON USES A FALSE NAME; AN ALIAS NAME. JON USES A FALSE LICENSE, I SEE A MAN'S NAME ON A LEASE. A MAN IS A MASON.(CEMENT, BUILDER, CONSTRUCTION).I CAN SEE A FALSE NAME ON A LEASE.

I CAN SEE JON IS NEAR MR LUNSFORD! I CAN SEE JESSICA'S REMAINS ARE FOUND NEAR MR LUNSFORD .(NEAR HER HOME). JON MURDER'S JESSICA, JESSICA DIES.

I CAN SEE JON CONFESSES A CRIME; A MURDER.I CAN SEE JON IS IN JAIL.

SEE? A SEER CAN SEE AND READ A MACRO CODE IN A NAME! AMEN = "ME A NAME AMEN."

AUTHOR: J.A.M. Channeled in Feb 2005 © copyright

JESSICA MARIE LUNSFORD SYNOPSIS

As you can see Jessica's name vibration tells a cameo of her short life. The acts, the scene's the story is all there by only using the letters of her name.

Again I have to say that there are no accidents, no random acts. The name itself say's see a macro code in a name.

In Jessica's name it say's I am near Jessica's soul; A seer's ear near Jessica's Soul. I am a seer near Jessica's soul.

It is time to see that the truth can be found within a name.

The name of the perpetrator can be found in the name. Again we can see entanglement with other soul entities. In Jessica's name alone her families criminal past are mentioned. Law enforcement in the acts is there; Even personal descriptions of Jessica's life is in her name. Jessica fondness and she rides a mare. animals

Not all names are tragic as in Jessica's case. Like Jesus has said: "a good name is better than a precious ointment."

It is a shame we can't get law enforcement to see past there five senses. The answers are right in front of us.

THE NAME ART OR SCIENCE OF THE AUTHOR

JOYCE ANN MORFORD

JOYCE ANN MORFORD CAME FROM D.N.A!

READ MY D.N.A. NAME (ART.) READY?

ME A NAME AMEN! MY NAME MADE ME!

ME: JOYCE ANN MORFORD, I READ MY NAME!

OM!

JOYCE A MONAD FROM ONE FORCE

JOYCE CAME FROM ONE FORCE!

CAME FROM ONE AND NONE!

FROM AN OAR OR ON= (LIGHT) OF MANY- FORM ONE! FROM NO FORM: COME A FORM!

A RAY OF ONE AND NONE! NEMO= NO MAN

AMON-RA MADE ME READ MY NAME! AMON-RA; MADE FROM (LIGHT)!

A NAME A MACRO CODE! A NAME A FORM!

JOYCE A DEACON (MESSNEGER, SERVANT) OF MERCY, (RIGHT HAND PILLAR) ON= (LIGHT); MON= (MOON GODDESS). JOYCE A FORMER DEACON OF AMORC (ANCIENT MYSTYCAL ORDER

OF ROSAE CRUCIS) JOYCE CAN MANCY= (PROPHET, TO PROPHECY) JOYCE CAN MANCY A NAME, FROM A NAME.

JOYCE A MONAD, (SPRITUAL WITNESS) JOYCE AND MONAD (SPIRIT) ARE ONE! JOYCE NEAR A MONAD- (SPIRIT) JOYCE NEAR MANY A MONAD (SPIRIT) JOYCE A RARE ONE! JOYCE CAN READ ONOMANCY= (DIVINATION BY THE LETTERS OF A NAME) JOYCE CAN DO ONOMANCY; JOYCE EAR NEAR MANY NECRO (DEAD) MONAD (SPIRIT) JOYCE CAN DO NECROMANCY! AN EAR NEAR MANY A MONAD= .(CLAIRAUDIENT)

JOYCE A FORM OF A CORONER= (public officer primary function is to investigate, by inquest, any death thought to be of other than natural causes.) JOYCE CAN DO NARCOMANCY (-SLEEP DIVINATION.) OR (MECONO-MANCY.)

JOYCE CAN DO NECROMANCY. (COMMUNICATE WITH DEAD SPIRITS)

JOYCE CAN DO AEROMANCY- (DIVINATION OF CLOUD SHAPES, APPEARENCES IN THE AIR.

JOYCE CAN DO CEROMANCY, (OBESERVATION OF CURRENTS, BRAIN WAVES) = TORRENT D.N.A= (PEER TO PEER NETWORK).

JOYCE CAN READ MANY AN OMEN (AUGURY), MANY AN OMEN IN A NAME.

JOYCE CAN RECORD A MODERN RECORD OF A NAME ON CD-ROM.

JOYCE A MONAD OF OAR OR ON (SPIRIT OF LIGHT)

DEAMON= (GENIUS=HOLY GUARDIAN ANGEL, TWIN SELF, HIGHER SELF) NEAR JOYCE, JOYCE AND DEAMON ARE ONE! JOYCE AND DEAMON FORM A CANON= (A REGULATION OR DOCTRINE DECREED BY A CHURCH COUNCIL) OF ONE! JOYCE= AN EAR NEAR AMON-RA (HIGHER SELF)

JOYCE CAN READ A NAME! JOYCE CAN READ AND DO NOMANCY= (THE ART OR PRACTICE OF Divining THE DESTINY OF PERSONS BY THE LETTERS WHICH FORMS THERE NAMES!) JOYCE CAN DO NOMANCY IN CAMEO= (A VIVID PORTRAYAL LITERARY) JOYCE CON NO ONE!

MY MA-MAE NEAR ME JOYCE. JOYCE AN EAR NEAR MA-MAE!(GRANDMOYHER)

JOYCE NEAR AN EAR: JOYCE NEAR AMEN RA, OR AMON RA JOYCE A DANCER, JOYCE DREAM OF RA, A JOY FOR MANY, A JOY FOR MANY YEARS. JOYCE DREAM OF D.N.A AND R.N.A.- JOYCE CAN FORM ONE D.N.A. ME A NAME JOYCE; JOYCE READ MANY A NAME. JOYCE ENJOY NOMANCY. JOYCE ENJOY ONOMANCY, JOYCE ENJOY NECROMANCY. DANCE FOR MORE JOY, MORE ROOM FOR JOYCE, MORE MONEY COME FOR JOYCE. MORE MONEY COME ON A MAY DAY, END OF MAY! ON A MOON DAY!

MAN-NA COME FROM AMON RA, MARY, MA, MAN-NA FROM AMON-RA-MAN-NA (DEW FROM HEAVEN) .

JOYCE CAN FORM A CORONA=(A FAINTLY COLORED LUMINOUS RING APPEARING TO SURROUND A CELESTIAL BODY, VISABLE THROUGH A HAZE OR THIN CLOUD, ESPECIALLY SUCH A RING AROUND THE MOON OR SUN, CAUSED BY DIFFRACTION OF LIGHT FROM SUSPENED MATTER IN THE INTERVENING MEDIUM, A CORONA OF THE SUN FROM AN ECLIPSE)

CORONA= (A GLOW FROM THE SUNS RAY FROM ECLIPSE OF THE MOON)

JOYCE RECORD MANY A NAME ON A CD-ROM, JOYCE READ MANY A NAME FROM A CD-ROM (AKASHIC RECORD)

JOYCE NO FROM DRACO- AN EAR NEAR DRACO.JOYCE NO DRACO= (DRAGON)

NO DEMON IN JOYCE. JOYCE NO RED DRACO (DRAGON) FROM DRACO, JOYCE CARE FOR AMON- RA, CARE FOR MA, OM, (Amen=, = Egyptian Sun God of life and reproduction (DNA) = The Sun, Son)A Man with a Rams Head ! JOYCE A RAM.

ROMANCE - A DEAR MAN ROMANCE JOYCE-JOYCE ROMANCE A DEAR MAN

FORM A DREAM, DREAM A FORM, READ MORE NO FEAR, ORDER A DREAM

ORDER MORE, MORE FOR JOYCE, MORE COME FOR JOYCE, JOYCE; ORDER MORE MONEY FOR ME!

MORE COME FROM AMON RA, FROM MA, FROM MANY A MONAD. (SPIRITS)

MANY A MONAD COME FOR JOYCE MORFORD, JOYCE A MARY, MARY A JOYCE, MARY AND JOYCE ARE ONE AMEN. AMON RA AND MA AND JOYCE ARE ONE, FROM OM, JOYCE A RAM

AMEN RA-MA MADE JOYCE, MADE MARY (Mary Magdalene)

A DEAR MAN NEAR ME

ME NEAR A DEAR MAN

A MAN READ MY NAME; A MAN NAMED DON FORCE, JOYCE READ A NAME, A MAN NAMED DON FORCE CAME FROM AMORC.C DON FORCE A DEACON FROM AMORC. JOYCE READ A NAME, A MAN NAMED JAROD.JOYCE READ A MAN NAMED ROY! JOYCE RE-READ A MAN NAMED ROY! ROY A MEAN MAN! NO MORE MEAN MEN FOR JOYCE! JOYCE CAN READ JOREN AND AN ANN.

ON MANY A DAY MY DEAR MAN NEAR ME- MADE MY DAY

MANY MEAN MEN, NO MORE MEAN MEN. JOYCE CAN DEFACE MANY MEAN MEN. JOYCE CAN DEFACE MANY A NAME CAN READ MANY A MEAN MEN NAME.

MANY MEN DAMN A MA, DAMN MANY A MAN

A MA AMEND

NAE (NO) MEAN MEN NEAR MARY

MANY MEAN MEN DENY MARY, DENY JOYCE

A DEAR MAN MADE MARY A MA (JESUS)

MARY MADE A DREAM, JOYCE MADE A DREAM

A RAM NEAR JOYCE- JOYCE A RAM, A DEAR MAN A RAM (Jeff rising sign Aries) NEAR JOYCE, JOYCE ANDA DEAR MAN ARE ONE, ARE FROM ONE D.N.A.

A DEAR MAN A RAM NEAR JOYCE, NEAR AN EAR, FOR A YEAR

MY DREAM, MADE ME; MADE MY NAME JOYCE MORFORD

A NAME MEAN ME, ME, A NAME AMEN

JOYCE NARD A DEAR MAN NEAR AN END (Nae Means NO orA MA MY MA NAME -MARYANN (Joyce's mothers name)

JOYCE DO MORE FOR A DEAR MAN, JOYCE AND A DEAR MAN ARE ONE

A DEAR MAN DO MORE FOR JOYCE

JOYCE CARE FOR MANY, JOYCE FOND OF MANY.

JOYCE FROM FRANCE; JOYCE FROM NORMANDY FRANCE. MARY FROM FRANCE, JOYCE READ CODE FROM FRANCE, FROM DNA, JOAN D ARC FROM FRANCE= (DOMREMY FRANCE) (past Life)

MEDOC= (RED BORDEAUX WINE FROM FRANCE)

MEDOC= ACRONYM= (MULTI POLE EXPANSION DEFINED ON ONE CENTER)

AMORC= ACRONYM= (ANICIENT MYSTIC ORDER OF ROSAE CRUCIS)

JORDEN= ACRONYM (POT OR VESSEL WITH A LARGE NECK USED BY ALCHEMISTS)

MONA (LISA) AND JOCONDE ARE ONE NAME. JOCONDE: A NAME FROM FRANCE. A MONA AND AMON ARE ONE!

OMO= ACRONYM= (ORACLE MEDIA OBJECT)

JOYCE CAN READ A CODON= A SEQUENCE OF THREE ADJACENT BASES ON A STRAND OF D.N.A. R.N.A. THAT PROVIDES GENETIC CODE INFORMATION FOR A PATICULAR AMINO ACID. JOYCE CAN READ FROM A D.N.A. CODON.

JOYCE MARRY A DEAR MAN, A RAM, ON A MAY DAY

A MAJOR DREAM COME FOR JOYCE

A CAR COME FOR JOYCE FROM MA

JOYCE MADE MONEY FROM A DREAM

JOYCE MADE MONEY FROM AN ACRONYM

JOYCE CAN FORM AN ACRONYM

JOYCE CAN FORM AN ACRONYM FOR MANY

OM= (The Movement of Solar Energy through the Chakras)

JOYCE DREAM OF A ROOM, A DOOR, A FACE OF RA, MA OM

A MOON RAY (Moon being receptive)

JOYCE CAN READ ANYONE; JOYCE CAN READ ANYONE FROM A NAME. COME FROM MOON (Feeling Empathic). A MOON, A MA NEAR JOYCE.

JOYCE CAN CRY NO MORE. ONLY JOY FOR JOYCE, FROM RA, FROM MA

DNA A CODE FROM RA, FROM MA, FROM D.N.A (OM) A NAME, A FORM OF DNA. JOYCE READ D.N.A AN EAR NEAR R.N.A.

JOYCE CAN DO A CAMEO OF A NAME - (A BRIEF VIVID PORTRAYAL OR DEPICTION, TO PORTRAY IN SHARP DELICATE RELIEF LITERARY COMPOSITION) A (LITERARY) CAMEO.

A NEON DYE= (A BRIGHT FLOURESCENT PURPLE) AN ELEMENT THAT GLOWS REDDISH ORANGE) IN JOYCE.

JOYCE READ A CODE, D.N.A, AN EAR NEAR R.N.A. JOYCE CAN READ A CODE, A CODON.

JOYCE CAN READ A FORM, JOYCE CAN READ ANY FORM! A CODE FROM RA, FROM MA

A DOOR FORM FOR JOYCE. JOYCE CAN READ D.N.A FOR MANY; JOYCE CAN READ A CODE FROM A NAME, JOYCE READ A NAME, A FORM OF D.N.A.

JOYCE CAN DECODE D.N.A- DECODE A NAME. A NAME CAN RECODE D.N.A!

JOYCE CAME FROM ONE FORM M&F (ACRONYM FOR FEMALE OR MALE energy), OR MORF (MALE OR FEMALE) ONE FORM.(androgynous) JOYCE READ A NAME ON A MONDAY= (MOONDAY) JOYCE READ MANY A NAME ON A MONDAY! ON ANY DAY!

MANY A DOOR FORM FOR JOYCE!

A NAME A CODE, A CANON D.N.A AND R.N.A (Change the Vibrations of Your Name, Your Code, Your DNA Through Meditation OM) (Positive thought or no thought) =OM

JOYCE CAN FORM A MODERN CANON OF ONE! A NAME A CANON. = (A LAW ON EARTH), JOYCE A FORM OF A MOON OM (Solar energy running through the Chakras)

NO MORE ROD FOR JOYCE (You control Life by controlling yourself, Throw down your Rod and it became a Serpent) No More Rod (it became Kundalini- OM (The movement of solar energy thru the Chakras- Kundalini)

JOYCE AN OCEAN, A MA

JOYCE NEAR AN OCEAN, A MER (MEANS WATER)

JOYCE CAN READ A RAY, RAY OF RA, RAY OF MA, A RAY OF MANY

(Spirit= The vital animating Force within living beings Incorporeal Consciousness that is spirit) JOYCE CAN READ A NOON RAY.

JOYCE A FORCE= (Electro magnetism= light= God is Light we are made in his image. Communicate with God, with Light) OM =D.N.A (GOD FORCE) (A LIGHT FORCE)

JOYCE AND FORAMEN ARE ONE= (Foramen definition: Amen Ra the Sun God ,(Solar Energy) Amen is connected to Jesus, Revelations 3:14 Jesus is called the AMEN. The person of Jesus and the person of Amen can be considered as one and the same. (9) In the Brain the opening of fornix is energy moving from one side to the other (8) Vault or tomb of Christ. The energy which is the Sun or solar energy moves out of the tomb into the conscious realm of perfection. It is for Amen, Christ Consciousness which is Within You!

Amen in the Bible is Jesus!!!

JOYCE MADE OR FORM AN ARC=OM= (A VESSEL THAT CARRIES THE DEVINE SEED AND SACRED KNOWLEDGE) JOYCE FORM AN ARC NEAR MARY, NEAR RA (north and South Poles- Male and female energy)

CODES COME EARLY IN MORN FOR JOYCE, FROM RA, FROM MA. JOYCE CAN READ A CARD, AN OMEN! = (PREDICTION) AN OMEN FORM FOR JOYCE. JOYCE READ AN OMEN AN ACE! JOYCE CAN READ AN OMEN IN A NAME!

JOYCE AN AEON OF ONE, AN AEON OF MA AND RA = (A DIVINE POWER OR NATURE, EMANATING FROM THE SUPREME BEING, AND PLAYING VARIOUS ROLES IN THE OPERATION OF THE UNIVERSE.)

AEON CARD= (corresponds to Aries-20th card in the tarot Judgment or Aeon, (Pluto in Aries) 20-JUDGEMENT CARD IN TAROT=VICTORY OF LIFE OVER DEATH. JOYCE; AN EAR NEAR A MOON RAY.= (SUBCONSCIOUS) RENEWAL; FRESH VIBRATIONS, HIGHLY SPIRITUAL.PRIME SOURCE FIRST SEED DIVINE SPARK OF LIFE.THAT WHICH ENDURES INDEFINATLEY.)

MACRO-CODE= (A code that contains several instructions, several sets of code within a code) D.N.A A NAME, A MACRO CODE IN A NAME! (MANY LAYERS OR CODES IN A NAME SAME AS D.N.A.) A MACRO CODE FROM D.N.A!

ODE= (A story, a song, a writing, a revealing in ideas one can Hear and Read) JOYCE AN EAR NEAR AN ODE, IN D.N.A, AN EAR IN R.N.A, JOYCE CAN CARRY AN ODE. JOYCE CAN FORM AN ODE ON CD-ROM!

ENAMOR= (means: love the source of all that is) JOYCE CAN ENAMOR A DEAR ONE. JOYCE CAN ENAMOR ONE FORCE.

JOYCE CAN READ A MACRO CODE FROM A NAME. JOYCE CAN DECODE A MACRO CODE FROM A NAME.

JOYCE FORM A YOD NEAR MOON, NEAR MA, NEAR RA, JOYCE FORM AN EAR NEAR AN ARC. JOYCE FORM A YOD= (IMPELLED TO CREATE)

MY D.N.A. AND R.N.A, FORM A YOD NEAR AMON-RA, NEAR MA, NEAR MARY

JOYCE AND JOAN D' ARC ARE ONE- JOAN OF ARC FROM FRANCE (CLAIRVOYANT)

JOYCE CAN READ JOAN OF ARC FROM FRANCE.

NO ROME FOR JOAN D'ARC.NO FEAR OF ROME, NO ROME FOR JOYCE, NO FEAR OF ROME, JOYCE DENY AN ERRANCY= (ACCEPTED DOCTINES OF THE CHURCH) OF ROME! NO ONE CAN DENY JOYCE CAN READ A NAME.

JOYCE READ A CODE FROM FRANCE, READ ACRONYM FROM FRANCE, JOYCE DECODE MANY AN ACRONYM FROM FRANCE, JOYCE DECODE MANY A NAME FROM FRANCE

ARCA=TOMB (ARK OF GOD) ARCA-DAY-E (ARC OF RA ARCADE)

AR= (THE PLOUGH= MEANS PENTAGRAM IN SUMERIAN)

ARC=TOMB, BUT ALSO MEANS A CHEST BOX, MONEY COFFIER IN LATIN, A CELL FOR CLOSE IMPRISONMENT=ARCANE= (BURIED AND CONCEALED,) JOYCE CAN READ AN ARCANE CODE! ARCEO=TO PROTECT KEEP SAFE, +OR HINDER, PREVENT), JOYCE CARE FOR MANY!

ARCEO- ALSO MEANS= (KING OF THE FOUR CORNERS OF THE EARTH)

A MAJOR CODE COME FOR JOYCE, JOYCE DECODE, A MAJOR CODE, A MAJOR MACRO CODE.

JOYCE FROM A MER RACE= AND A MARY RACE; AND A JOAN RACE; A RA, AND MA RACE. A DRACO RACE (dragon) ONE RACE, A RACE OF ONE!

JOYCE ADORE MANY, ADORE RA, ADORE MA

JOYCE CAN READ A CARD, AN OMEN, JOYCE CAN READ AN OMEN FROM A CARD, JOYCE READ AN ACE ON A CARD! AN ACE AN OMEN!

JOYCE CAN FORM NOMANCY.COM

MANY ORDER A CD-ROM FROM NOMANCY .COM

JOYCE MADE MORE MONEY FROM NOMANCY.COM

JOYCE EARN FAME FROM NOMANCY .COM

MANY ORDER A NAME ON CD-ROM.

MANY READ A MODERN CANON FROM NOMANCY.COM

C A MAJOR DREAM COME FOR JOYCE.

ACRONYM-MER=MAXIMUM EFFICIENT RATE

ACRONYM-ROM= READ ONLY MEMORY- JOYCE READ FROM A ROM-D.N.A A ROM, (D.N.A AKASHA RECORD) ACCESS DATA THAT CANNOT BE CHANGED.MEMORY HARDWARE THAT ALLOWS ACCESS TO PERMANTLY STORED DATA.

JOYCE A CDROM=(REMOVING AN OUTGOING TRANSMISSION FROM THE ORIGINAL SOURCE, A DISK THAT HOLDS LARGE AMOUNTS OF INFORMATION THAT CAN BE READ BY A COMPUTER BUT CANNOT BE CHANGED (D.N.A) AKASHA RECORD-CRYSTALLINE) D.N.A, R.N.A= (TRANSMISSION OF GENETIC INFORMATION)

JOYCE CAN READ A NAME, A NAME D.N.A, A NAME R.N.A A CODE ON A CDROM, A NAME A CODE ON CDROM (A RECORD=IN THE ASTRAL IS AKASHA-ASTRAL LIGHT RECORD)

OAR= (LIGHT) JOYCE CAN READ a RECORD OF OAR- (LIGHT) A RECORD OF (LIGHT).

EMF= ELECTRO MAGNETIC FIELD

ARM=ACRONYM=AUTOMATIC REMOTE MODE (AUTOMATIC WRITING, AWARENESS OF INTERNAL PROCESSES) = (Self Awareness)

YOD=ACRONYM=Y OUTPUT DATA=TRANSMISSION MEDIUM

RAM=ACRONYM= RANDOM ACCESS MEMORY= A MEMORY DEVICE IN WHICH INFORMATION CAN BE ACCESSED IN ANY ORDER

JAM=ACRONYM= JAMES THE BROTHER OF JESUS=JAM=JESUS AND ME

JAM= JOINING ALL MINDS

DRAM=DYNAMIC RANDOM ACCESS
MEMORY=MEMORY THAT ALLOWS ACCESS TO

ANY MEMORY LOCATION WITHOUT RESTRICTIONS

RAD=RECEIVE ANTENNA DIVERSITY=A THREE ANTENNA SYSTEM, POLARITY DIVERSITY RECEPTION, USES BOTH HORIZONTAL AND VERTICAL +POLARAZATION ANTENNAS TO RECEIVE THE SIGNAL.

REMOTE ANTENNA DRIVE=REMOTE VIEWING

RA=RADIUM- CHEMICAL TABLE-ATOMIC NUMBER 88- NUMBER OF PROTONS/ELECTRONS 88-ATOMIC MASS 226, NUMBER OF NEUTRONS 136-OBTAINED FROM URANIUM ORES.7 ENERGY LEVELS-COLOR SILVERISH-CUBIC

NA=SODIUM-ATOMIC NUMBER-11-PROTONS/ELECTRON 11-NEUTRONS 12-ALKALI METAL CRYSTAL-CUBIC-SILVERY-ENERGY LEVEL3 (2-8-1) NAME ORIGIN SODA USES MEDICINE, AGRICULTURE

FR-FRANCIUM-ATOMIC NUMBER-87-PROTONS/ELECTRON-87-NEUTRONS 136-ALIKALI METAL CRYSTAL- CUBIC-COLOR UNKNOWN- ENERGY LEVEL7 (2-8-18-32-18-8-1) NAME ORIGIN FRANCE-USE UNKNOWN

CA=CALCIUM- ATOMIC NUMBER 20-ENERGY LEVELS 2-8-8-2-COLOR SILVER-CUBIC

CR=CHROMIUM-ATOMIC NUMBER 24-PROTONS/ELECTRONS 24-NEUTRONS 28-

ENERGY LEVELS 4-92-8-13-1) COLOR GRAY-A TRANSITION METAL.CUBIC

Y=YITTRIUM-ATOMIC NUMBER 39-ATOMIC MASS-88.90585-BOILING POINT 3337.0-PROTONS/ELECTRONS 39-NEUTRONS 50-TRANSITION METAL- HEXAGONAL COLOR SILVERY-ENERGY LEVELS 5(2-8-18-9-2) USES COLOR TV'S. RADAR

MN=MAGNESIUM-ATOMIC NUMBER 25, PROTONS/ELECTRONS 25-NEUTRONS 30-TRANSITION METAL-SILVERISH-GRAYISH-USES-MAGNET- BATTERIES, STEEL, CERAMICS ENERGY LEVELS 4(2-8-13-2

FE=IRON-ATOMIC NUMBER-26, ATOMIC MASS 55.845 PROTON/ELECTRONS 26-NEUTRONS 30-TRANSITION METAL-ENERGY LEVELS 4 (2-8-14-2) CARRIES OXYGEN IN THE BLOOD IRON ORE-CUBIC

CO=COBALT- ATOMIC NUMBER27-PROTONS/ELECTRONS 27-NEUTRONS 32-TRANSITION METAL-HEXAGONAL-ENERGY LEVELS 4 (4-8-15-2) COLOR SILVER- USES-MAGNETS

MO=MOLYBDENUM-ATOMIC NUMBER 42-PROTONS/ELECTRONS-42-NEUTRONS-54-TRANSITION METAL-ENERGY LEVELS 5 (2-8-18-13-1) ORIGIN -LEAD.USES-AIR CRAFT MISSLES.CRYSTAL-CUBIC

CD=CADMIUM-ATOMIC NUMBER-48-PROTONS/ELECTRONS-48-NEUTRONS/64-TRANSITION METAL-HEXAGONAL-ENERGY LEVELS 5 (2-8-18-18-2) ANCIENT NAME-CALAMINE-USES-NICKEL- CADMIUM BATTERIES

RE=RENIUM-ATOMIC NUMBER-75-PROTONS/ELECTRONS-75-NEUTRONS-111-TRANSITION METAL-HEXAGONAL-ENERGY LEVEL 6 (2-8-18-32-13-2) SILVERISH-USES FILAMENTS FOR SPECTOGRAPHS (EQUIPPED TO PHOTOGRAPHS)

RF=RUTHERFORDIUM-ATOMIC NUMBER 104-PROTONS/ELECTRONS-104-NEUTRONS-157-ENERGY LEVELS 7 (2-8-18-32-32-10-2)

C=CARBON-ATOMIC NUMBER 6-PROTONS/ELECTRON-6-NEUTRONS-6-NON METAL-HEXAGONAL-BLACK -ENERGY LEVELS-2 (2-4)-COAL-BURNING WITH INSIFFICIENT OXYGEN

N=NITROGEN-ATOMIC NUMBER-7-PROTONS/ELECTRONS-7-NEUTRONS-7- NON METAL-HEXAGONAL—COLORLESS-ENERGY LEVELS 2 (2-5) FORMS MOST OF ATMOSPHERE-OBTAINED FROM LIQUID AIR

0=OXYGEN-ATOMIC NUMBER 8-PROTONS/ELECTRONS-8-NEUTRONS-8-CRYSTAL STRUCTURE-CUBIC-COLORLESS-ENERGY LEVELS 2-(2-6) SUPPORTS LIFE FROM LIQUID AIR

F=FLOURINE-ATOMIC NUMBER-9-PROTONS/ELECTRONS-NEUTRONS-10-CRYSTAL-CUBIC-COLOR-GREENISH-ENERGY LEVELS 2(2-7) USES REFIGERANT-(LATIN WORD FLOW)

NE-NEON-PROTONS/ELECTRONS-10-NEUTRONS-10-NOBLE GAS-CUBIC-ENERGY LEVELS 2 (2-8) USES LIGHTENING-OBTAINED FROM LIQUID AIR

AR=ATOMIC NUMBER-18-PROTONS/ELECTRONS-18-NEUTRONS-22 ARGON-PROTONS/ELECTRONS-18-NEUTRONS-22-NOBLE GAS-CUBIC-COLORLESS-ENERGY LEVEL 3 (2-8-8)-USES LIGHTENING-OBTAINED FROM AIR

RN=RADON-ATOMIC NUMBER-86-PROTONS/ELECTRONS-86-NEUTRONS-136-NOBLE GAS-COLORLESS-ENERGY LEVELS 6 (2-8-18-32-18-8) FROM RADIUM-TREATMENT OF CANCER

CE=CERIUM=ATOMIC NUMBER-58-PROTONS/ELECTRONS-58-NEUTRONS-82-RARE EARTH-CUBIC-GRAY-ENERGY LEVELS 6 (2-8-18-20-8-2) ORIGIN ASTROID CERES-USES HEAT RESISTANT ALLOYS-OBTAINED FROM MONAZITE

ND=NYODIMIUM-ATOMIC NUMBER-60-PROTON/ELECTRONS-60-NEUTRONS-84-CRYSTAL STRUCTURE-HEXAGONAL-COLOR-SILVERY-ENERGY LEVELS 6-(2-8-18-22-8-2-) GREEK WORD-NEOS-2) (DIDYMOS-TWIN*) USES - COLORING GLASS, AND CERAMICS-INFRARED RADIATION FILTERING OBTAINED FROM-ELECTROLYSIS SALTS

DY=-DYSPROSIUM-ATOMIC NUMBER-66-PROTONS/ELECTRONS-66-NEUTRONS-97-CRYSTAL STRUCTURE-CUBIT-ENERGY LEVELS 6-(2-8-18-28-8-2) NAME ORIGIN GREEK-DYSPOSITOS (HARD TO GET AT) USES-NUCLEUR REACTORS

ER=ERBIUM=ATOMIC NUMBER-68-PROTONS/ELECTRONS-68-NEUTRONS 99-RARE EARTH-CRYSTAL STRUCTURE-HEXAGONAL-COLOR GRAYISH-ENERGY LEVEL 6(2-8-18-30-8-2) USES CERAMICS

AC=ACTINIUM-ATOMIC NUMBER-89-PROTONS/ELECTRONS-89-NEUTRONS-138-RARE EARTH-CRYSTAL-CUBIC-COLOR SILVERY-ENERGY LEVELS-7 (2-8-18-32-18-9-2) FROM GREEK WORD AKTINOS-(RAY)* EXTREMELY RARE!

AM=AMERICIUM=ATOMIC NUMBER-95-PROTONS/ELECTRONS-95-NEUTRONS-148-RARE EARTH-CRYSTAL-HEXAGONAL-ENERGY LEVELS 7(2-8-18-32-25-8-2) USES- SMOKE DETECTERS

CM=CURIUM=ATOMIC NUMBER 96-PROTONS/ELECTRONS-96-NEUTRONS-151-RARE EARTH-CRYSTAL STRUCTURE-UNKNOWN-COLOR UNKNOWN-ENERGY LEVELS 7-(2-8-18-32-25-9-2) NO USES KNOWN

CF-CALIFORNIUM-ATOMIC NUMBER-98-PROTONS/ELECTRONS-98-NEUTRONS-153-RARE

EARTH-CRYSTAL-UNKNOWN-COLOR-UNKNOWN-ENERGY LEVELS-7 (2-8-18-32-28-8-2)

FM=FERBIUM=ATOMIC NUMBER-100-PROTONS/ELECTRONS-100-NEUTRONS-157-CRYSTAL-UNKNOWN-COLOR-UNKNOWN-ENERGY LEVELS 7 (2-8-18-32-30-8-2)

MD=MENDELVIUM-ATOMIC NUMBER-101-PROTONS/ELECTRON-101-NEUTRONS-157-RARE EARTH-CRYSTAL STRUCTURE UNKNOWN COLOR UNKNOWN-ENERGY LEVELS 7(2-8-18-32-31-8-2)

NO=NOBELIUM=ATOMIC NUMBER-102-PROTON/ELECTRONS-102-NEUTRONS-157-RARE EARTH CRYSTAL AND COLOR UNKNOWN-ENERGY LEVELS 7(2-8-18-32-32-8-2) USES UNKNOWN

JOYCE FORM ONE D.N.A, JOYCE AND DAEMON ARE ONE, ONE MONAD OF OAR OR ON! (LIGHT) MA AND RA COME FOR JOYCE FROM COMA! JOYCE COME FROM A RARE RACE; A RACE OF ONE!

JOYCE ANN MORFORD SYNOPSIS

When I obtained Christ Consciousness I was in a state of oneness; the total loss of my EGO (false self).I was walking the Earth with my mind in a Quantum state still attached to my physical body. In this state there is only Joy, Peace, Oneness and Belonging to everyone and everything. I was able to read the acts the scenes and the story in all Human names including my own.

To the average reader; when reading the channel they may or may not understand it.

I am doing the channeling as an observer with an objective mind and reality, with no judgement, no emotions.

I am only using the letters of the name which my consciousness unscrambles to reveal my reality with this Name and all other Names on Earth; whether it is a person a place a thing any name describes the acts the scenes the story of itself.

I actually mirror the holder of the Name- the consciousness of the entity that holds the Name.

As you read the first portion of the name art it reveals my Self Realization of whom and what I actually am. I know I am not my body. I know I am a light force in the body. All my chakras are open from the kundalini rising out through the top of my head; my crown chakra. I have achieved cosmic consciousness. I am emanating a light of the one original Monad, (My Soul) I am Eternal!

All my chakras are open and I have access internally to my DNA, my Akashic records. I pull out spiritual gifts I had in other life times. Which are psychic gifts, Clairvoyant, Clairaudient, Clairsentience, and Clair-cognizant. In one of my lifetimes I was a High Priestess in Egypt. The reference to Ra is a portion of that lifetime in Egypt. The reference to Edgar Cayce points to the lifetime in Egypt we were associated when he was Ra-Ta. I had great intuitive abilities and I was a healer. Amon-Ra refers to that lifetime. I have a connection with the Great Pyramid in Giza knowing it is a consciousness ascension device, a tool per say to raise it. My consciousness went through the initiation in this life time, a remembrance from my life in Egypt.

In another life time I was burned at the stake in Salem Mass for my clairvoyant abilities and my gift of Nomancy. In my name is "Joyce a Joan of Arc" Referring to me being burned at the stake. I reincarnated with the same group of Souls (my accusers) in this lifetime to heal all karmic issues, unfortunately those karmic issues were not resolved with the Souls I am entangled with in this lifetime. I came here this lifetime to raise the consciousness and vibration of humanity using my gifts. I awakened to my true self, while they stayed asleep. (meaning being ruled by their Ego, their sleeping false self, and only using the logical conscious mind, instead of being whole and listening to the intuition first, the innate}.

I started to read everyone's name that I knew, to see if it is indeed a new canon for humanity. I see this truth every day.

In my name art you will see: Joyce a form of a coroner. (by this it means by using the name of a person or the names of the person's involved I can see the criminal acts. I investigate the deaths that seem to come from unnatural causes; in such a case as someone trying to hide a murder, kidnapping, all crime scenes. I find the truth using the person's names; nothing hides in a name.

The down fall of this state of consciousness is that most of humanity is ignorant to the fact that the "Christ Consciousness" is raised within you. I was misunderstood due to their ignorance. When I am doing name arts my ego is not there. I am in an altered state of consciousness; a Quantum state of consciousness. This is the mind of oneness, no ego; no separation and nothing is hidden. To the humans in a Duality Consciousness the EGO rules!

If a person has been hiding things they do not want others to know the EGO (the false self) gets angry, defensive, afraid, desperate to hide the truth of one's own self. Your name reveals many of the things you have done in your life both good and bad. Your name can reveal past lifetimes, sickness to health that may incur. The duality consciousness labels everything as either good or bad. In the Christ Mind all is oneness and there is no judgement. There is only Joy, Peace, Oneness and belonging.

So the Christ Consciousness is available to every human being on earth. It is up to the person to raise this Consciousness. This requires inner work and reprogramming all the lies that have been told for eons.

In Christ Consciousness there is no disease of the mind, body and soul. This is how peace on Earth will be established. Humanity only needs to wake up to their reality and know that all is just temporary here on Earth. Most of humanity does not realize the purpose of their life is to experience Self-realization in the body. Most of humanity believes they are here to go to school, to get a job. To get married, to have children, so they can buy all there material items that they hang on to. The problem with humanity is they lack any spiritual-ism within the self. They nurture their mind and there body and they leave spirit completely out.

To me Humanity was done a great dis-service from acts from long ago, where the powers that be could take advantage of them for their own selfish ways. In actuality it's not their fault per-say. The mind can be brainwashed which they took complete advantage of.

My name art states exactly what I am here for and what I am to accomplish. I am here to raise the vibration of humanity with spiritual truth that they are unaware even exists.

When you have an EGO death and resurrection in the body; you are truly born anew; for your consciousness is now vibrating at a higher and faster frequency. So the life you had in duality consciousness is washed away and all those, mistakes, flaws, and situations you are not very proud of, all that is washed away. You're emotional or pain body is born anew with love and oneness; for yourself and also everyone and everything.

The choice is yours. If you are not a seeker of the truth then nothing changes and your reality stays the same. It is the seeker whom is looking for "Why am I here" Who am I" "What's it all for". Everything is in a quantum state, and it is the asking that changes your reality instead of outside in- it is changing from the inside out.

This work that I am doing is my gift to humanity. It is not to attack anyone.

My name art is my experience here on earth NOW!

In my name art you will read about a higher being name RA. Ra can be channeled at any time. Most of the children of sorrow send out thought messages that can be picked up by a higher being. A higher being can also see how far on your spiritual path you are by the colors emitting from your aura; your energy fields. This is how I was contacted. Carla Ruekoff is the original channel of Ra and the "Law of One" a must read for everyone.

My name art also tells when I met and married my husband at the time.

It touches on his mother being mean; and also that I A Seer can see... FREEDOM FROM A MEAN MA! AMEN! That is another story in itself that I won't go into here at this time.

My name art reveals my psychic gifts, clairvoyance, clairaudience, clairsentinance, Clair-cognizant abilities.

I also saw my future, which at this time I cannot reveal here.

My work here is about the Oneness of us all!

SECTION 3

CONCLUSION A METAPHYSICAL TREATISE

Definition of Metaphysical= Relating to a reality beyond what is acceptable to the senses. Of or relating to things that are thought to exist but cannot be seen.

Some would call it supernatural

In this Metaphysical Treatise I have shown you what is visible beyond the 5 senses. A higher state of consciousness is developed from the inside-out in meditation; stilling the mind.

What I perceive in a higher vibration is much different than what another perceives in a lower vibration; meaning they are only in touch with the 5 senses.

In conclusion no one knows what is happening to you on the inside of yourself. Your perception and reality on the inside will not be the same as another, it cannot be viewed from someone looking at you from the outside. Higher Consciousness is experienced from the inside; any spiritual gifts that you obtain will be from the works that you do from the inside out. Maybe you become a healer, maybe you become an Intuitive empath like myself with a special gift or tool that you use or do to access your gift. There are many talents and gifts stored in your D.N.A. that will manifest for your higher calling in this life. Everyone is growing at their own rate.

Christ Consciousness is this: Your consciousness is in a quantum state while still attached to the physical body. When consciousness is raised to a quantum state you perceive your reality as oneness and wholeness. There is only Love, Joy, Peace and Oneness. You are walking the Earth just as Jesus the Christ did.

For in that state of consciousness there is only LOVE; there is no duality, no polarity, and no separateness, no ego.

You create a new self with a higher awareness where the animal Soul is washed away; and you become a spiritual Soul with a higher vibrational frequency. You literally are born anew!

This metaphysical treatise I leave for humanity for I raised my consciousness to this state. It is obtainable through meditation and the asking for such a state by going within you. Quote: Jesus said; the kingdom of heaven is within you and this is true. Jesus said: These things I do you can do also and more!

Peace on Earth requires this state of oneness. Humanity has been duped by the churches' and the Elite of the Earth for Eons to have Power over the People.

Remember the church did not want you to go with in, for they knew the truth. The church kept you coming to them for your own salvation and asking you to pay for it. To me Humanity should be on the brink of Spiritual Rage.

Isn't it time for you to take back your power and your God given spiritual gifts.

You are a co-creator in this universe = (One Verse). You are not small at all! You are equal to the All!

A person having a dharma life is very unlikely to turn within his/her self for any kind of guidance, they will be too busy enjoying the material world. In itself it is not a bad thing; it just doesn't give you a chance for self-realization in the body until after your death.

A person having a more karmic life, whether it is their own, or whether they volunteered to be in karmic conditions helping the soul of the other redeem its karmic balance in order to make the turn to a more positive vibration. Under extreme karmic condition's when one has volunteered to help the other soul's karma to be balanced; there is the chance that the other parties will not make the turn to a more positive vibration healing all karmic debts. You volunteered to help the soul on the other side before you came here; out of unconditional love for that soul; you may endure great suffering (as was my case in this lifetime) in which there is nowhere else to turn but deep within yourself for comfort, for asking the reasons why this is happening, for love, and this is where the magic is. This is where you find yourself. This is where self- realization happens; and you find that you are not your body; you are not your personality because you have had an ego death in the body. You find out that you are a light with multidimensional communicating abilities with the whole universe. You find out that you have had many lifetimes, many memories which are stored in D.N.A. D.N.A stores all cellular memories of all your lifetimes it is call the Akashic records. So every lifetime you are creating your record for the next lifetime.

This information has been hidden from you from long -long ago. The priests from ancient Egypt had the knowledge that was given to them by more advanced beings. These beings are called - 9 the Ennead, whom I believe Ra is connected to. A higher race of what we would call extraterrestrial beings) they transmitted their message using telepathy to a being named Ra- Ta. Ra-TA left the Egyptians when they started distorting the truth. The royal elite took control over the people and distorted the "Law of One" from what is initially intended for. More information can be found on Carla Ruekoff "Law of One "website.

9 the Ennead truly exists as a higher intelligence that has never been human. They have also been referred as the "Elohim".

My own contact with what I call; the blue race, came from telepathic communication within myself. The pouring out of unconditional love from these higher beings is phenomenal. These beings are not to be feared. They are here to uplift humanity in a positive way. The want you to know them and that your also apart of them.

All this information came from within me, from a higher part of me. The Monad, the Higher self, the Over Soul of me. This is why meditation is important in seeking the truth. When you are born, you are born with a veil that splits the conscious mind from the subconscious mind. Before incarnating you are whole and know everything. The test here on Earth is the test of Love and the ways of love. It's a test if you can find the God inside you. You are a co-creator; you are a piece of God.

I am not my body, I am a light; a spirit having a physical experience to find the truth here on Earth. This means you are a light also animating your physical body here on Earth; acting out the dream on the physical stage of life. Whether that truth is karma or dharma, for we meet these here on the physical plane to balance the positive and negative polarity. The opportunity to awaken to a higher a truth; "Me" a Name Amen, my reality and experiences were from my name and the names of entanglement with others names. In fact you could say: Change your name-Change your Life! So in and of itself I am you and you are me; staring back at each other having a physical reality. I am ONE! Here is a Key: All letters are vibration. The letters in your name are interchangeable and so creates acts, scenes and events in your life. Here is proof of seeing Quantum Entanglement in this physical reality we call life. It answers the question of how and why events occur within our lives. Like Jesus the Master of Love stated "A good name is truly better than a precious ointment" With this knowledge maybe an awakening will occur in that knowing maybe negative traits and scenes ahead of time we can choose the right hand path and bypass the negative outcome.

Peace on Earth is knowledge to raise our vibration as humanity to Love as a whole collective Equal to each other yet an individualized light.

Namaste

DEDICATION

This book is dedicated to all humanity

THE DARE

Raising your consciousness to a higher state is given to the seeker. The seeker need only to ask with crystal clear intent in order for the transformation to begin. Everything is entangled in a quantum state. Consciousness is entangled in a quantum state. If you are not a seeker then nothing in your consciousness will change; it will stay the same until you seek for the truth.

Christ Consciousness or Quantum Consciousness is a state of Love and Oneness; that is the state of consciousness Jesus was in when he was talking to his followers. His PERCEPTION was very different from the Duality Minds of the people (Souls) were in.

My experience in the state of being in Christ Consciousness has forever change my view of life. I am only able to explain my reality in words; but that is like trying to tell a blind man what the color blue looks like.

Living amongst the ego duality mindset while you are in a Quantum state of Oneness mindset is quite the nightmare. With all the hatred around me; meaning the ones who I thought were my loved ones, was in reality trying to destroy me little by little due toothier ignorance; and of course due to there wanting to hide some very incriminating acts.Some of the things that were done to me was quite horrific and had I not been in a Christ Conscious state I would have been destroyed, Emotionally, Mentally and Physically.

I am here to say there is more than one Consciousness on Earth available here, if you so choose to access it. In a higher state of Consciousness you become a self-realized being. You become the Watcher, the Observer and your EGO is not present.

What is the Dare you ask?

The dare is to make a present day movie of the 2 different states of Consciousness, and the 2 different perception's. Duality is a low energy life force steeped in animalistic behavior operating from the 3 lower energy centers of the body. The Oneness state of Consciousness starts with the Heart Chakra energy centers and upward and they are all opened so the animal nature is burned away. Now the animal nature is replaced by the Spiritual nature of man. This enables the Spiritual man to be able to communicate with his Soul.

The nightmare of it all is that the conditioning of the masses has left people in ignorance believing Jesus our savior is returning; When in reality

Christ Consciousness is within you to be risen. In the duality mind set you are viewed as CRAZY, because of their ignorance of anything else but their programmed existence.

NOTES

Law of One Material

Author Channel Carla Reukoff

Quote: Ra, I am Ra- All names can be taken for their vibrational characteristics.

Ra- I am Ra- The law of one Though beyond the limitations of name (Very Important here, Ra says beyond the limitations of a name) All things are one No polarity, no disharmony, no righteousness, but only identity all is one, therefore Love/Light, Light/Love.

Ra, I am Ra- If one wishes to have power over an entity, it is an aide to know that entities name.

If one wishes no power over an entity but wishes to collect that entity into the very heart of one's own being, it is well to forget the name.

Edgar Cayce Companion-page172-Evolution of mind and matter

There is a name to each soul.

Edgar Cayce.org, Biblical reference, Book of life: Akashic records- Names

The Bible: Quote biblical reference, Jesus had said

Ecclesiastes 7:1- A good name is better than a precious ointment.

Made in the USA
Monee, IL
06 November 2026

68085973R00104